RECREATIONAL PILOT

FAA WRITTEN EXAM

FIRST (1989-1991) EDITION

by Irvin N. Gleim, Ph.D., CFII

ABOUT THE AUTHOR

Irvin N. Gleim earned his private pilot certificate in 1965 from the Institute of Aviation at the University of Illinois, where he subsequently received his Ph.D. He is a commercial pilot and flight instructor (instrument) with instrument, multiengine, and seaplane ratings, and is a member of the Aircraft Owners and Pilots Association, American Bonanza Society, Civil Air Patrol, Experimental Aircraft Association, and Seaplane Pilots Association. He is also author of many pilot training books including flight maneuvers and handbooks for the recreational, private, instrument, commercial, and flight instructor certificates/ratings, and study guides for the FAA written tests for the recreational, private, instrument, and commercial pilot.

Dr. Gleim has also written articles for professional accounting and business law journals, and is the author of the most widely used review manuals for the CIA exam (Certified Internal Auditor), the CMA exam (Certificate in Management Accounting), and the CPA exam (Certified Public Accountant). He is Professor Emeritus, Fisher School of Accounting at the University of Florida, and is a CIA, CMA, and CPA.

Aviation Publications, Inc.
P.O. Box 12848 • University Station
Gainesville, Florida 32604
(904) 375-0772

Library of Congress Catalog Card No. 89-85120
ISBN 0-917539-23-0

Copyright © 1989 by Aviation Publications, Inc.

ALL RIGHTS RESERVED. No part of this material may be reproduced in any form whatsoever without express written permission from Aviation Publications, Inc.

ABBREVIATIONS USED IN *RECREATONAL PILOT FAA WRITTEN EXAM*

The first lines of the answer explanations contain citations to authoritative sources of the answers. Most of these publications are obtainable from the FAA. These citations are abbreviated as noted below:

AC	Advisory Circular	AWS	Aviation Weather Services
ACL	Aeronautical Chart Legend	FAR	Federal Aviation Regulations
AFD	Airport/Facility Directory	FTH	Flight Training Handbook
AFNA	Aerodynamics for Naval Aviators	NTSB	National Transportation Safety Board Regulations
AIM	Airman's Information Manual		
AvW	Aviation Weather	PHAK	Pilot's Handbook of Aeronautical Knowledge

---

HELP !!

This is the First Edition, designed specifically for potential recreational pilots. It will be revised biennially. Please send any corrections and suggestions for subsequent editions to the author, c/o Aviation Publications, Inc. The last page in this book has been reserved for you to make comments and suggestions. It can be torn out and mailed to Aviation Publications, Inc.

Also, please bring these books to the attention of flight instructors, fixed base operators, and others interested in flying. Wide distribution of this series of books and increased interest in flying depend on your assistance and good word. Thank you.

NOTE: ANSWER DISCREPANCIES and ERRATA SHEETS

Our answers have been carefully researched and reviewed. Inevitably, there will be differences with competitor books and even the FAA. If necessary, we will develop an ERRATA SHEET for *RECREATIONAL PILOT FAA WRITTEN EXAM*. Please write us about any discrepancies. We will respond to all inquiries.

Answers are either A, B, C, or D. Answer 5 means no correct answer and we believe the FAA is accepting any answer as correct. Relatedly, when we believe the FAA is accepting one answer as correct when one or more answers may be correct, or as correct, we indicate the FAA preferred answer as correct.

IF FOUND, please notify and arrange return to owner. This written test book is important for the owner's preparation for the Federal Aviation Administration Written Test for the Recreational Pilot Certificate. Thank you.

Pilot's Name _____

Address _____

City/State/Zip _____

Telephone (____) _____

Additional copies of *Recreational Pilot FAA Written Exam* are available from

Aviation Publications, Inc.
P.O. Box 12848 • University Station
Gainesville, Florida 32604
(904) 375-0772

The price is $9.95. Orders must be prepaid and are shipped postpaid; i.e., we pay the postage. Florida residents must add 6% sales tax.

Aviation Publications, Inc. guarantees an immediate, complete refund on all mail orders if a resalable book is returned within 30 days.

ALSO AVAILABLE FROM AVIATION PUBLICATIONS, INC.
See pages 208-210 for additional information and an order form.

Handbooks and Flight Maneuvers

RECREATIONAL PILOT FLIGHT MANEUVERS
PRIVATE PILOT HANDBOOK
PRIVATE PILOT FLIGHT MANEUVERS
INSTRUMENT PILOT FLIGHT MANEUVERS AND HANDBOOK
MULTIENGINE AND SEAPLANE FLIGHT MANEUVERS AND HANDBOOK
COMMERCIAL PILOT AND FLIGHT INSTRUCTOR FLIGHT MANEUVERS AND HANDBOOK

Written Exam Books

PRIVATE PILOT FAA WRITTEN EXAM
INSTRUMENT PILOT FAA WRITTEN EXAM
COMMERCIAL PILOT FAA WRITTEN EXAM
FUNDAMENTALS OF INSTRUCTING FAA WRITTEN EXAM
FLIGHT INSTRUCTOR/GROUND INSTRUCTOR FAA WRITTEN EXAM

REVIEWERS AND CONTRIBUTORS

Craig Delgato, CFI, B.S. in aviation management, Florida Institute of Technology, is a flight instructor and charter pilot at Gulf Atlantic Airways in Gainesville, Florida. He is a commercial pilot with instrument and multiengine ratings. Mr. Delegato reviewed the entire text, contributing to its technical accuracy.

Windy A. Kemp, B.S.Acc., University of Florida, assisted in the coordination of proofreading and production staffs. She will be joining a national public accounting firm in January 1990.

Patricia L. McGhee, B.A. University of Wisconsin-Madison, has 19 years of editing and production experience in scientific and technical publications. She reviewed the entire text and assisted with the coordination of its production.

John F. Rebstock, CIA, is a graduate of the School of Accounting at the University of Florida. He has passed the CPA exam and is a CMA candidate. Mr. Rebstock prepared the page layout, reviewed the entire text, and coordinated its production.

Gustave W. Schwartz, ATP, CFII, is Adjunct Professor at San Diego Mesa Community College in San Diego, California, and is an active flight instructor in the San Diego area. He reviewed the entire text and provided many useful suggestions.

Douglas E. Sims, CFI, is a flight instructor in single-engine and multi-engine airplanes. He reviewed the entire text, contributing to its technical accuracy.

The many FAA employees who helped, in person or by telephone, primarily in Gainesville, Florida; Orlando, Florida; Oklahoma City, Oklahoma; and Washington, DC.

The many CFIs, pilots, and student pilots who have provided comments and suggestions about all of my books during the past 8 years.

A PERSONAL THANKS

This manual would not have been possible without the extraordinary efforts and dedication of Ann Finnicum and Susan Young Burnett, who typed the entire manuscript and all revisions, as well as prepared the camera-ready pages. Ms. Burnett is also responsible for developing the typographic design utilized in this book.

The author also appreciates the proofreading assistance of Sandra Beasley, Debbie Durkin, Robert Francis, Appie Graham, Michael Kohl, Andy Mason, Marcia Miller, Leslie O'Donnell, Ketan Patel, Evan Rothman, Mary Ann Vorce, and Marie Wilker, and the production assistance of Darius Cauthen, Donald Dehne, Andy Mason, Ketan Patel, Cristina Shaw, Mary Ann Vorce, and Chris Yost.

Finally, I appreciate the encouragement, support, and tolerance of my family throughout this project.

TABLE OF CONTENTS

	Page
Preface	vi
CHAPTER 1 • The FAA Written Test	1
CHAPTER 2 • Introduction to Airplanes and Aerodynamics	15
CHAPTER 3 • Airplane Performance	29
CHAPTER 4 • Airplane Instruments, Engines, and Systems	49
CHAPTER 5 • Airports and Air Traffic Control	73
CHAPTER 6 • Weight and Balance	87
CHAPTER 7 • Weather	101
CHAPTER 8 • Federal Aviation Regulations	137
CHAPTER 9 • Navigational Publications and Sectional Charts	171
CHAPTER 10 • Flight Physiology and Flight Operations	185
FAA Listing of Subject Matter Knowledge Codes	195
Cross-References to the FAA Written Test Question Numbers	202
Index	213

CAUTION: The FAA issues new written test books in the Spring, and they usually expire September 1, two years later. Thus for about 6 months after a new written test book is issued, FAA written test examiners can test you with either the old FAA edition or the new FAA edition. Talk to your written test examiner to determine which edition (s)he will use. The dates of our editions are on the cover. Call with any questions.

PREFACE

The primary purpose of this book is to provide you with the easiest, fastest, and least expensive means of passing the recreational pilot (airplane) FAA written test. We have
1. Reproduced each of the 445 FAA test questions that can possibly be used on your written test.
2. Reordered the questions into 105 logical topics.
3. Organized the 105 topics into 9 chapters.
4. Explained the answer immediately to the right of each question.
5. Provided an easy-to-study outline of exactly what you need to know (and no more) at the beginning of each chapter.

Accordingly, you can thoroughly prepare for the FAA written test by
1. Studying the brief outlines at the beginning of each chapter.
2. Answering the question on the left side of each page while covering up the answer explanations on the right side of each page.
3. Reading the answer explanation for each question that you answer incorrectly or have difficulty with.

The secondary purpose of this study aid is to introduce RECREATIONAL PILOT FLIGHT MANEUVERS.

RECREATIONAL PILOT FLIGHT MANEUVERS is a practical description, with numerous illustrations and diagrams, of how "to fly" an airplane. In outline format, it is organized and written to assist you in meeting the new FAA Practical Test Standards (reproduced by topic in each chapter). The Practical Test Standards are the basis on which your FAA flight test will be conducted.

Additionally, this book will introduce our entire series of pilot training texts, which use the same presentation method: outlines, illustrations, questions, answer explanations, indexes, etc.

Most books create additional work for the user. In contrast, this written test book facilitates your effort. It is easy to use. The outline format, type styles, and spacing are designed to improve readability. Concepts are often presented as phrases rather than as complete sentences. Also note that we have located charts and figures next to the questions, not in appendices at the back of the book as in the *FAA Test Question Book*. This saves you time and aggravation during study. A sectional chart legend appears inside the front cover. The inside and outside of the back cover contain two sectional chart examples which are the basis of questions from Chapter Nine.

Read Chapter 1, "The FAA Written Test," carefully. Also recognize that this study manual is concerned with airplane flight training, rather than balloon, glider, or helicopter training. I am confident this manual will facilitate speedy completion of your written test. I also wish you the very best as you complete your recreational pilot certificate, in subseqent flying, and in obtaining additional ratings and certificates.

Enjoy Flying -- Safely!
Irvin N. Gleim
September 15, 1989

CHAPTER ONE
THE FAA WRITTEN TEST

```
What Is a Recreational Pilot Certificate? ............................................. 2
Requirements to Obtain a Recreational Pilot Certificate ............................ 2
FAA Written Test ....................................................................... 4
How to Prepare for the Written Test .................................................. 5
When to Take the Written Test ....................................................... 6
Where to Take the Written Test ...................................................... 6
Authorization to Take the Written Test ............................................... 7
Format of the Written Test ............................................................. 7
Examination Procedures ............................................................... 7
What to Take to the FAA Written Test ............................................... 11
FAA Written Test Instructions ....................................................... 11
How the FAA Notifies You of Your Written Test Score ............................. 11
If You Receive Less than a 70% Grade .............................................. 12
FAA Question Numbers ............................................................... 13
```

The beginning of this chapter provides an overview of the process to obtain a recreational pilot's certificate. The remainder of the chapter explains the content and procedure of the FAA written test. Learning to fly and getting a pilot's certificate are fun. Begin today!

This is one of two related books for recreational pilots. The other book is *RECREATIONAL PILOT FLIGHT MANEUVERS*, which is also in outline format. It is a complete text and reference of flight maneuvers for recreational pilots and student pilots. Included are diagrams for all the maneuvers. The FAA Practical Test Standards (PTSs) are also reprinted.

Sample Recreational Pilot Certificate

Front   Back

WHAT IS A RECREATIONAL PILOT CERTIFICATE?

A recreational pilot certificate is much like an ordinary driver's license. A recreational pilot certificate will allow you to fly an airplane and carry one passenger and baggage, although not for compensation or hire. However, operating expenses may be shared with your passenger. The certificate, which is a piece of paper similar to a driver's license, is sent to you by the FAA upon satisfactory completion of your training program, a written examination, and a flight test. A sample certificate is reproduced at the bottom of page 1.

REQUIREMENTS TO OBTAIN A RECREATIONAL PILOT CERTIFICATE

A. **Obtain an FAA medical certificate.**

   1. You must undergo a routine medical examination which may only be administered by FAA-designated doctors who are called FAA medical examiners.

   2. To obtain this medical certificate, the applicant must be at least 16 years old and be fluent in English.

   3. The medical certificate necessary for a recreational pilot certificate is called a third-class medical. It is valid for 2 years and expires on the last day of the month issued (when another medical examination is required).

   4. Even if you have a physical handicap, medical certificates can be issued in many cases. Operating limitations may be imposed depending upon the nature of the disability.

   5. Your flight instructor or fixed base operator (FBO) will be able to recommend an aviation medical examiner.

      a. *FBO* is an airport business that gives flight lessons, sells aviation fuel, repairs airplanes, etc.

      b. Also, the FAA publishes a directory that lists all authorized aviation medical examiners by name and address. Copies of this directory are kept at all FAA District Offices, Air Traffic Control facilities, and Flight Service Stations.

B. **The medical certificate will function as your student pilot certificate once it is signed (endorsed) on the back by your flight instructor.**

   1. Alternatively, a separate student pilot certificate can be obtained from a General Aviation or Flight Standards District Office.

   2. Note that the back of the student pilot certificate must be signed by your flight instructor prior to solo flight (flying by yourself).

   3. The front and back of a sample FAA medical certificate/student pilot license are reproduced on the next page.

   4. The only substantive difference between a regular medical certificate and a medical certificate/student pilot license is that the back of the medical certificate/student pilot license provides for flight instructor endorsements.

Chapter 1: The FAA Written Test 3

Front **Back**

C. **Pass a written test with a score of 70% or better.** Most FAA written tests are administered by FAA designated examiners. This test is administered at most General Aviation District Offices (GADOs) and at some airport FBOs. The test consists of multiple-choice questions selected from the airplane-related questions among the 500 questions in the *FAA Recreational Pilot Question Book* (the balance of 54 questions are for balloons, helicopters, etc.). Each of the FAA's airplane questions are reproduced in this book with complete explanations to the right of each question. One duplicate question is noted, but not reprinted. The questions test the following 9 topics:

1. Introduction to Airplanes and Aerodynamics
2. Airplane Performance
3. Airplane Instruments, Engines, and Systems
4. Airports and Air Traffic Control
5. Weight and Balance
6. Weather
7. Federal Aviation Regulations
8. Navigational Publications and Sectional Charts
9. Flight Physiology and Flight Operations

DUPLICATE QUESTIONS IN THE FAA WRITTEN TEST BOOK: There is one duplicate question which is noted in the second line of the answer explanation and in the cross-reference table on page 203. The 500 FAA questions consist of 54 nonairplane question, 445 airplane questions in this book, and one duplicate question.

You are required to satisfactorily complete a ground instruction or home study course prior to taking the written test. Page 207 contains a standard authorization form that can be completed, signed by a flight instructor or ground instructor, torn out, and taken to the examination site.

D. **Obtain at least 30 hours of flight instruction and solo flight time, including**

1. 15 hours of flight instruction from an authorized flight instructor, including at least

    a. Two hours away from the airport, including 3 landings at an airport at least 25 nautical miles (NM) away.

    b. Three hours at night, including 10 takeoffs and landings for applicants seeking night flying privileges.

    c. Two hours in airplanes in preparation for the recreational pilot practical test within 60 days prior to that test.

2. 15 hours of solo flight time in an airplane.

E. **Successfully complete a flight test which will be given as a "final exam" by an FAA inspector or designated examiner.** The flight test will be conducted as specified in the FAA's PTSs.

1. FAA inspectors are FAA employees and do not charge for their services.

2. FAA designated examiners are proficient, experienced flight instructors and pilots who are authorized by the FAA to conduct flight tests and administer FAA written tests. They do charge a fee.

3. The FAA's PTSs are reproduced in *RECREATIONAL PILOT FLIGHT MANEUVERS*.

FAA WRITTEN TEST

This written test book is designed to help you prepare for and successfully take the FAA written test for the recreational pilot certificate.

A. The remainder of this chapter explains the FAA written test procedures.

B. All of the 445 questions in the *FAA Recreational Pilot Question Book* (FAA-T-8080-14) that are applicable to airplanes have been grouped into the following nine categories, which are the titles of Chapters 2 through 10:

        Chapter 2 • Introduction to Airplanes and Aerodynamics
        Chapter 3 • Airplane Performance
        Chapter 4 • Airplane Instruments, Engines, and Systems
        Chapter 5 • Airports and Air Traffic Control
        Chapter 6 • Weight and Balance
        Chapter 7 • Weather
        Chapter 8 • Federal Aviation Regulations
        Chapter 9 • Navigational Publications and Sectional Charts
        Chapter 10 • Flight Physiology and Flight Operations

Note that in the official FAA Question Book, the questions are not grouped together by topic. We have unscrambled them in this book.

C. Within each of the chapters listed above, questions relating to the same subtopic (e.g., duration of medical certificates, stalls, carburetor icing, etc.) are grouped together to facilitate your study program. Each subtopic is called a module.

D. To the right of each question is the correct answer and one or two paragraphs explaining the correct answer and the incorrect answers.

Chapter 1: The FAA Written Test   5

E. Each chapter begins with an outline of the material tested on the FAA written test. The outlines in this part of the book are very brief and have only one purpose: to help you pass the FAA written test for recreational pilots.

1. CAUTION: The sole purpose of this book is to expedite your passing the FAA written test for the recreational pilot certificate. Accordingly, all extraneous material (i.e., not directly tested on the FAA written test) is omitted even though much more information and knowledge is necessary to fly safely. This additional material is presented in RECREATIONAL PILOT FLIGHT MANEUVERS.

Follow the suggestions on the following pages and you will have no trouble passing the written test the first time you take it.

HOW TO PREPARE FOR THE WRITTEN TEST

A. Begin by carefully reading the rest of this chapter. You should have a complete understanding of the examination process prior to beginning to study for it. This knowledge will make your studying more efficient.

B. After you have spent an hour studying this chapter, set up a study schedule, including a target date for taking the test.

1. Do not let the study process drag on and become discouraging, i.e., the quicker the better.

C. As you move from module to module and chapter to chapter, you may need further explanation of certain topics. We hope you graduate from this written test book to PRIVATE PILOT HANDBOOK and RECREATIONAL PILOT FLIGHT MANEUVERS. Along with this book, these books contain all the information in the FAA's Pilot's Handbook of Aeronautical Knowledge, Flight Training Handbook, Aviation Weather, FAA Recreational Pilot Practical Test Standards, and information relevant to student pilots and recreational pilots from over 30 other FAA books, booklets, brochures, etc.

1. If you are going to use PRIVATE PILOT HANDBOOK, obtain it now. Note that it has the same chapter organization as this book. See pages 208 to 210 for additional information and an order form.

2. Many concepts discussed in conjunction with the FAA written test questions will need further clarification. These can be obtained by referring to the index at the back of PRIVATE PILOT HANDBOOK.

D. Note that this written test book (in contrast to other question and answer books) reorders the FAA questions by topic. Thus, some are repetitive. Accordingly, you should not work question after question (i.e., waste time and effort) if you are already conversant with a topic and the type of questions asked.

E. Keep track of your work!!! As you complete a module in Chapters 2 through 10, grade yourself with an A, B, C, or ? next to the module title on the first page of each chapter.

1. The A, B, C, or ? is a self-evaluation of your comprehension of the material in that module, and your ability to answer the questions. We suggest the following guide:

    A means a good understanding
    B means a fair understanding
    C means a shaky understanding
    ? means to ask your CFI or others about the material and/or questions, and read the pertinent sections in RECREATIONAL PILOT FLIGHT MANEUVERS and PRIVATE PILOT HANDBOOK.

2. The titles of the modules in each chapter are listed on the first page of each chapter. The number of FAA questions that cover the information in each module is indicated in parentheses next to each module title. The two numbers following the parentheses are the page numbers on which the outline and the questions for that particular module begin, respectively.

3. By following the self-evaluation procedure described on the previous page, you will be able to see quickly (by looking at the first page of Chapters 2 through 10) how much studying you have done (and how much remains) and how well you have done.

4. This procedure will also facilitate review. You can spend more time on the modules you had difficulty with.

F. Work through each of Chapters 2 through 10.

1. Begin by studying the outlines slowly and carefully.

2. Cover the answer explanations on the right side of each page with your hand or a piece of paper while you answer the multiple-choice questions.

   a. Study the answer explanation for each question that you answer incorrectly, do not understand, or have difficulty with.

   b. Remember, it is very important to the learning (and remembering) process that you honestly commit yourself to an answer. If you are wrong, your memory will be reinforced by having discovered your error. Therefore, it is crucial to cover the answer column and select your choice from the alternatives before reading the answer explanation.

WHEN TO TAKE THE WRITTEN TEST

A. Take the written test within the next 30 days.

1. Get the test behind you.

B. Take your practical test within 24 months of your written test,

1. Or you will have to retake your written test.

WHERE TO TAKE THE WRITTEN TEST

A. Most FAA written tests are administered by FAA designated examiners. Written tests are also administered by some FAA

1. Flight Standards District Offices (FSDOs).
2. Air Carrier District Offices (ACDOs).
3. Flight Service Stations (FSSs).

B. Ask your instructor or call a nearby airport to inquire about the nearest FAA facility administering written tests.

C. There is no charge to take the written test at a FAA facility.

D. Call to verify that the FAA facility you are considering administers written tests (some do not).

E. Also, many FBOs, in conjunction with a FAA-designated examiner, administer written tests for a nominal fee, e.g., $20. Check with your flight or ground instructor.

AUTHORIZATION TO TAKE THE WRITTEN TEST

The FAA requires applicants for the recreational pilot certificate to *have logged ground instruction from an authorized instructor, or present evidence showing that they have satisfactorily completed a course of instruction or home study in at least the following areas of aeronautical knowledge appropriate to the category of aircraft for which a rating is sought:*

1. *The Federal Aviation Regulations applicable to recreational pilot privileges, limitations, and flight operations, accident reporting requirements of the National Transportation Safety Board, and the use of the Airman's Information Manual and the FAA Advisory Circulars.*
2. *VFR navigation, using pilotage, dead reckoning, and radio aids.*
3. *The recognition of critical weather situations from the ground and during flight, and the procurement and use of aeronautical weather reports and forecasts.*
4. *The safe and efficient operation of airplanes, including high density airport operations, collision avoidance precautions, and radio communication procedures.*

For your convenience, a standard authorization form is reproduced on page 207 which can be easily completed, signed by a flight instructor or ground instructor, and torn out for you to take to the written test site.

FORMAT OF THE WRITTEN TEST

The FAA's recreational pilot test for airplanes consists of 50 multiple-choice questions taken from the 445 questions that appear in the next 9 chapters. You will have 4 hours to complete the test, which is plenty of time.

Note that the questions on the FAA written test are exactly the same ones that are reproduced in this volume. If you study the next 9 chapters, including all the questions and answers, YOU SHOULD BE ASSURED OF PASSING YOUR FAA WRITTEN TEST.

EXAMINATION PROCEDURES

When you arrive at the FAA facility or your FBO, register for the test by completing the information section of the answer sheet. The answer sheet is labeled "Airman Written Test Application" (hereafter referred to as the answer sheet). A sample answer sheet is presented on page 8. It is important that you complete all items.

The "Test No." at the upper right comes from the "Question Selection Sheet." A question selection sheet similar to the one you will be given is presented on page 9. Note that your question sheet will have only 50, not 60, items and will have a 2-, not 3-column format. The question selection sheet tells you which of the airplane-related questions you are to answer.

Chapter 1: The FAA Written Test

DEPARTMENT OF TRANSPORTATION — FEDERAL AVIATION ADMINISTRATION

AIRMAN WRITTEN TEST APPLICATION

DATE OF TEST	TITLE OF TEST	TEST NO.
MONTH 02 / DAY 15 / YEAR 90	RECREATIONAL PILOT AIRPLANE	237498

PLEASE PRINT ONE LETTER IN EACH SPACE—LEAVE A BLANK SPACE AFTER EACH NAME

NAME (LAST, FIRST, MIDDLE): Doe John R.

DATE OF BIRTH: MONTH 03 / DAY 16 / YEAR 53

MAILING ADDRESS NO. AND STREET, APT. #, P.O. BOX, OR RURAL ROUTE: 2536 N. Main Street

CITY, TOWN OR POST OFFICE, AND STATE: Louisville, Kentucky
ZIP CODE: 40502

DESCRIPTION: HEIGHT 70" / WEIGHT 155 / HAIR Br / EYES Br

BIRTHPLACE (City and State, or foreign country): Buechel, Ky.
CITIZENSHIP: U.S.A.
SOCIAL SECURITY NO.: 411-23-2677
IF A SOCIAL SECURITY NUMBER HAS NEVER BEEN ISSUED CHECK THIS BLOCK ☐

Is this a retest? ☑ No ☐ Yes, date of last test
Have you taken or are you taking an FAA approved course for this test? ☐ No ☑ Yes (if "yes" give details below)

Graduation date:
NAME OF SCHOOL: Kenair Flying School
CITY AND STATE: Louisville, Kentucky

CERTIFICATION: I CERTIFY that all of the statements made in this application are true, complete, and correct to the best of my knowledge and belief and are made in good faith. Signature _John R. Doe_

— — DO NOT WRITE IN THIS BLOCK — — FOR USE OF FAA OFFICE ONLY — —
Applicant's identity established by.

CARD A				CARD B			FIELD OFFICE DESIGNATION
CATEGORY	TEST NUMBER	TAKE NO.	SECTIONS 1 2 3 4 5 6 7	EXPIRATION MONTH DAY YEAR	CERTIFICATED SCHOOL NUMBER	MECH EXP DATE BY SECTION 1 2 3	ID	SIGNATURE of FAA Representative

INSTRUCTIONS FOR MARKING THE ANSWER SHEET. Completely darken only one circle for each question. DO NOT USE (X) OR (✓). Use black lead pencil furnished by examiner. To make corrections, open answer sheet so erasure marks will not show on page 2. Then erase incorrect response on page 4. On page 2 (copy) mark the incorrect response with a slash (/). Questions are arranged in VERTICAL sequence as indicated by the arrows.

Sample

Chapter 1: The FAA Written Test 9

1. The 445 questions are in a separate book entitled *Recreational Pilot Question Book*. The number on the front of the book of questions should be the same number as on the top of the question selection sheet. For example, the number on the sample question selection sheet below is FAA-T-8080-14. Bring any discrepancy to the attention of the person in charge of the test.

DEPARTMENT OF TRANSPORTATION · FEDERAL AVIATION ADMINISTRATION

QUESTION SELECTION SHEET

Use only with Question Book FAA-T-8080-14

TITLE	TEST NO.
RECREATIONAL PILOT - AIRPLANE	237498

NAME _____

NOTE: IT IS PERMISSIBLE TO MARK ON THIS SHEET.

On Answer Sheet For Item No.	Answer Question Number	On Answer Sheet For Item No.	Answer Question Number	On Answer Sheet For Item No.	Answer Question Number
1	201	21	414	41	604
2	214	22	420	42	612
3	220	23	425	43	652
4	226	24	433	44	665
5	229	25	440	45	680
6	236	26	456	46	699
7	241	27	462	47	700
8	253	28	470	48	718
9	258	29	478	49	724
10	262	30	481	50	750
11	274	31	484	51	775
12	295	32	498	52	799
13	288	33	501	53	803
14	309	34	506	54	845
15	318	35	515	55	853
16	366	36	522	56	868
17	376	37	528	57	875
18	382	38	534	58	905
19	392	39	552	59	951
20	403	40	560	60	976

For Official Use Only

After you complete the information section (top) of the answer sheet, you will be given

1. The answer sheet (returned to you by the person in charge of the test after (s)he has checked to see that you have completed the application portion properly).
2. The question selection sheet.
3. *Recreational Pilot Question Book* (FAA-T-8080-14).
4. Pencils and scratch paper (otherwise use the back of the question selection sheet as scratch paper).
5. Plastic overlays.

The plastic overlays are to put over pages on which you must mark, e.g., routes between airports on sectional chart excerpts. The actual book of test questions is reused continually, and the FAA asks that you make no marks directly on it. Thus, plastic overlays permit you to write on the plastic (which is erasable), rather than on the pages in the exam question book itself.

A. Before starting the test, place your answer sheet off to the side where it will be out of your way, and forget about it until after you have completed taking the test.

B. When you complete a question, write the answer you have selected on your question selection sheet, not on your answer sheet. Then check the question number to ensure that you have answered the correct question.

C. Answer the easy questions first. If you have trouble with a question or want to temporarily skip a question, place an "X" next to the question number on the question selection sheet and move on to the next question.
  1. If you are unsure of an answer, write the answer you believe is correct on your question selection sheet and place a "?" next to the question number.
  2. After working through all 50 questions, go back and answer all of the questions you marked with an "X" or a "?".

D. As you read each question, ignore (or cover up) the alternative answer selections. Most people taking tests skim a question to "get on" to the answers.
  1. When reading the question, pay particular attention to the requirement, i.e., EXACTLY WHAT IS REQUIRED?
  2. If you do not completely understand the requirements, you cannot answer the question correctly.
  3. Then go back and use the data in the question and any related sectional chart excerpts, diagrams, etc. to solve the question.
  4. Select the best answer.
  5. Mark the answer on the question selection sheet.

E. Watch your time. You have almost 5 minutes per question (50 questions in 240 minutes). Most candidates finish in 1 to 2 hours.

Chapter 1: The FAA Written Test   11

F. After you have answered all 50 questions, get up, stretch, and walk around for a few moments. Then, carefully transfer your answers to your answer sheet.

   1. Before turning in your answer sheet, check to see that the first 50 answers are marked and that only one answer is marked for each question.

   2. If you change an answer, make sure that erasures are complete.

G. Note that the FAA answer sheet is numbered to 150 instead of 50. This is because the answer sheet is designed for all FAA exams. You are only to mark answers 1 through 50 (the number of questions on the FAA Recreational Pilot Written Test).

H. Finally, remember that you only need a 70% to pass. 70% of 50 questions is 35 questions. Thus, you can miss 15 and still pass.

   1. While you should seriously attempt every question, do not become obsessed with making a perfect score. If you do, you may become rattled if you have trouble with a question. Also, you will begin to spend too much time on each question and become fatigued before you complete the test.

WHAT TO TAKE TO THE FAA WRITTEN TEST

   1. Navigational plotter.

   2. Pocket calculator if you are familiar with it and have used it before (no instructional material for the calculator is allowed).

   3. Authorization to take the examination (see page 7).

   4. Note: Paper and pencils are supplied at the examination site.

   5. Picture identification of yourself.

FAA WRITTEN TEST INSTRUCTIONS

The page of general instructions that appears in the test book is reproduced on page 14. Study the instructions now and reread them when you take the written test.

HOW THE FAA NOTIFIES YOU OF YOUR WRITTEN TEST SCORE

A. The results of your written test will be sent to you from the FAA Aeronautical Center in Oklahoma City where it is graded, approximately 2 weeks after you take the exam.

B. Your written test score will be sent to you on the "Airman Written Test Report" FAA Form AC 8080-2, which is reproduced on the following page.

   1. Note that you will receive only one grade as illustrated.

   2. The expiration date is the date by which you must take your FAA flight test.

   3. It lists the FAA subject matter knowledge codes of the questions you missed, so you can review the topics you missed prior to your flight test.

C. Use the FAA list of subject matter knowledge codes on pages 195 to 201 to determine which topics you had difficulty with.

1. Look them over and explain your review procedure to your CFI so (s)he can sign your Airman Written Test Report to the effect that (s)he reviewed the deficient areas and found you competent in them.

D. Keep your Airman Written Test Report in a safe place, as you must submit it to the FAA examiner when you take your final flight test.

DO NOT DESTROY THIS TEST REPORT This Test Report must be presented for retesting or certification.	DEPARTMENT OF TRANSPORTATION - FEDERAL AVIATION ADMINISTRATION **AIRMAN WRITTEN TEST REPORT (RIS: AC 8080-2)**	1466 41 SSN 434-80-6677

TEST		GRADES BY SECTION							FAA OFFICE NO.	TEST DATE	EXPIRATION DATE
TAKE NO.	TITLE*	1	2	3	4	5	6	7			
01	RA	77							SO 07	02-18-88	02-28-90

EXPIRATION DATE (Last day of month)

*See codes on reverse side:

MECHANICS ONLY - EXPIRATION DATE CODES
The first character designates the month; the second and third characters, the year. January through September as shown by numbers 1 through 9; October as "O"; November as "N"; December as "D".

LAST NAME, FIRST MIDDLE
FLANAGAN WILLIAM PATRICK JR
4720 NW 39TH ST
GAINESVILLE FL   32601

EXAMPLES:
Month (June) — 6 75
Year (1975) ——
Month (December) — D 75
Year (1975) ——

1   B05 C01 C10 D06 D26 D27 H03 J10 J11 K18 K25 Q02 Q12 R11

When applicable, an authorized instructor may complete and sign this statement:
I HAVE GIVEN THIS APPLICANT ADDITIONAL INSTRUCTION IN EACH OF THE SUBJECT AREAS FAILED AND CONSIDER THE APPLICANT COMPETENT TO PASS THE TEST.
LAST_____ INITIAL_____ CERTIFICATE NO._____ TYPE_____ INSTRUCTOR'S SIGNATURE_____
    INSTRUCTOR'S NAME (Print)
FRAUDULENT ALTERATION OF THIS FORM BY ANY PERSON IS A BASIS FOR SUSPENSION OR REVOCATION OF ANY CERTIFICATES OR RATINGS HELD BY THAT PERSON.   ISSUED BY: ADMINISTRATOR
AC Form 8080-2 (10-83)   FEDERAL AVIATION ADMINISTRATION

IF YOU RECEIVE LESS THAN A 70% GRADE

A. If you fail the written test (highly unlikely if you follow the previous instructions), you may retake it after 30 days.

1. You can retake the test sooner than 30 days (after a first failure only) if you obtain a written note from a Certificated Flight Instructor (CFI) that you have received the necessary ground instruction to retake the test.

2. As a practical matter, it takes 2 weeks to get the test to Oklahoma City for grading and the test score returned to you, so the actual delay is only about 2 weeks.

Chapter 1: The FAA Written Test

B. Upon retaking the test, everything is the same except you must submit your Airman Written Test Report indicating the previous failure to the examiner.

C. The pass rate on the Recreational Pilot Written Test will probably be about 80%, i.e., 2 out of 10 fail the test initially. Reasons for failure include

1. Failure to study the material tested, i.e., the outlines at the beginning of Chapters 2 through 10 of this book.

2. Failure to practice working the FAA exam questions under test conditions (all of the FAA questions on airplanes appear in Chapters 2 through 10 of this book).

3. Poor examination procedure, such as

    a. Not reading questions and understanding the requirements.

    b. Failure to complete the answer sheet correctly, e.g., getting answers out of sequence.

FAA QUESTION NUMBERS

A. The questions in the FAA written test book are numbered 0001 to 0500. The FAA questions appear to be presented randomly.

1. We have reorganized and renumbered the FAA questions into chapters and modules.

2. The FAA question number is presented in the first line of the explanation of each answer.

B. Pages 202 through 205 contain a list of the FAA questions numbers 0001 to 0500 with cross references to the chapters and question numbers in this book.

1. For example, question 0001 is coded 4-25, which means it is found in this book as question 25 in Chapter 4.

2. Remember, the questions in this book have been reorganized by topic rather than randomly as they appear in the FAA Question Book.

3. Note that, although 0001 to 0500 implies 500 questions, only 445 apply to airplanes.

    a. The remaining 54 questions (e.g., those on helicopters, gliders, balloons, etc.) have been omitted from this book.

        1) There is no cross-reference to our book if they are not airplane questions.

    b. Also note that only 445 questions appear in this book as the FAA has one duplicate question (which is labeled as such).

    c. In summary, 445 in this book + 54 nonairplane + 1 duplicate = 500 questions.

With this overview of exam requirements, you are ready to begin the easy-to-study outlines and rearranged questions with answers to build your knowledge and confidence and PASS THE FAA's RECREATIONAL PILOT WRITTEN TEST.

Although no quantifiable data have been collected, feedback we receive from users of our texts indicate that our written exam books lessen anxiety, improve FAA test scores, and build knowledge. Studying for each test thus becomes a useful step toward advanced certificates and ratings.

Chapter 1: The FAA Written Test

GENERAL INSTRUCTIONS

MAXIMUM TIME ALLOWED FOR TEST: 4 HOURS

TEST MATERIALS

Materials to be used with this question book when used for airman testing:

1. AC Form 8080-3, Airman Written Test Application, which includes the answer sheet.
2. Question selection sheet which identifies the questions to be answered.
3. Plastic overlay sheet which can be placed over performance charts for plotting purposes.

TAKING THE TEST

1. Read the instructions on page 1 of AC Form 8080-3, and complete page 4 of the form.
2. The question numbers in this question book are numbered consecutively beginning with number 0001. Refer to the question selection sheet to determine which questions to answer.
3. For each item on the answer sheet, find the appropriate question in the question book.
4. Mark your answer in the space provided for that item on the answer sheet.
5. Certain answer sheets may contain responses listed as 1, 2, 3, and 4 and should be interpreted as A, B, C, and D respectively.
6. Read each question carefully and avoid hasty assumptions. Do not answer until you understand the question. Do not spend too much time on any one question. Answer all of the questions that you readily know and then reconsider those you find difficult. Be careful to make necessary conversions when working with temperatures, speeds, and distances.

If a regulation, chart, or operations procedure is changed after this question book is printed, you will receive credit for the affected question until the next question book revision.

Comments regarding this publication should be directed to:

U.S. Department of Transportation
Federal Aviation Administration
Aviation Standards National Field Office
Examinations Standards Branch
Operations Standards Section, AVN-131
P.O. Box 25082
Oklahoma City, OK 73125

THE MINIMUM PASSING GRADE IS 70.

WARNING

§61.37 Written tests: Cheating or other unauthorized conduct.

(a) Except as authorized by the Administrator, no person may-
(1) Copy, or intentionally remove, a written test under this part;
(2) Give to another, or receive from another, any part or copy of that test;
(3) Give help on that test to, or receive help on that test from, any person during the period that test is being given;
(4) Take any part of that test in behalf of another person;
(5) Use any material or aid during the period that test is being given; or
(6) Intentionally cause, assist, or participate in any act prohibited by this paragraph.

(b) No person whom the Administrator finds to have committed an act prohibited by paragraph (a) of this section is eligible for any airman or ground instructor certificate or rating, or to take any test therefore, under this chapter for a period of 1 year after the date of that act. In addition, the commission of that act is a basis for suspending or revoking any airman or ground instructor certificate or rating held by that person.

CHAPTER TWO
INTRODUCTION TO AIRPLANES AND AERODYNAMICS

Flaps and Rudder	*(3 questions)*	15, 19
Aerodynamic Forces	*(2 questions)*	15, 20
Angle of Attack	*(4 questions)*	16, 20
Stalls and Spins	*(2 questions)*	16, 21
Frost	*(3 questions)*	16, 22
Ground Effect	*(4 questions)*	16, 23
Airplane Turn	*(1 question)*	17, 24
Airplane Stability	*(5 questions)*	17, 24
Torque and P-Factor	*(3 questions)*	17, 26
Load Factor	*(7 questions)*	18, 26

This chapter contains outlines of major concepts tested, all FAA test questions and answers regarding the basics of aerodynamics, and an explanation of each answer. The subtopics or modules within this chapter are listed above, followed in parentheses by the number of questions from the FAA written test pertaining to that particular module. The two numbers following the parentheses are the page numbers on which the outline and questions begin for that module.

CAUTION: Recall that the sole purpose of this book is to expedite your passing the FAA written test for the recreational pilot certificate. Accordingly, all extraneous material (i.e., topics or regulations not directly tested on the FAA written test) is omitted, even though much more information and knowledge are necessary to fly safely. This additional material is presented in *RECREATIONAL PILOT FLIGHT MANEUVERS* and *PRIVATE PILOT HANDBOOK*, available from Aviation Publications, Inc. See pages 208 to 210 for more information and an order form.

FLAPS AND RUDDER (3 questions)

1. One of the main functions of **flaps** during the approach and landing is to increase wing lift, which allows an increase in the angle of descent without increasing airspeed.

2. The **rudder** is used to control the yaw about the airplane's vertical axis.

AERODYNAMIC FORCES (2 questions)

1. The four aerodynamic forces acting on an airplane during flight are
   a. **Lift**, the upward-acting force;
   b. **Weight** (or gravity), the downward-acting force;
   c. **Thrust**, the forward-acting force; and
   d. **Drag**, the rearward-acting force.

2. These forces are at equilibrium when the airplane is in unaccelerated straight-and-level flight:

$$\text{Lift} = \text{Weight}$$
$$\text{Thrust} = \text{Drag}$$

ANGLE OF ATTACK (4 questions)

1. The *angle of attack* is the angle between the wing chord line and the direction of the relative wind.
    a. The *wing chord line* is an imaginary straight line through the wing from the leading edge to the trailing edge of the wing.
    b. The *relative wind* is the direction of airflow relative to the airplane wing moving through the air.
2. The angle of attack at which a wing stalls **remains constant irrespective** of weight, airplane loading, etc.

STALLS AND SPINS (2 questions)

1. An airplane can be stalled at any airspeed in any flight attitude. **A stall results** whenever the critical angle of attack is exceeded.
2. An airplane in a given configuration will stall at the same indicated airspeed regardless of altitude because the airspeed indicator is directly related to air density.
3. An airplane spins when one wing is less stalled than the other wing.
    a. **To enter a spin**, an airplane must always first be stalled.

FROST (3 questions)

1. **Frost forms** when the temperature of the collecting surface is at or below the dewpoint of the adjacent air, and the dewpoint is below freezing.
    a. The water vapor **sublimates** directly as ice crystals on the wing surface.
2. Frost on wings disrupts the smooth airflow over the airfoil by causing early airflow separation from the wing. This
    a. **Decreases lift**, and
    b. Causes friction and **increases drag**.
3. Frost may make it difficult or impossible for an airplane to take off.
4. Frost should be removed before attempting to take off.

GROUND EFFECT (4 questions)

1. Airplanes are affected by ground effect when they are less than one-half the length of the airplane's wingspan above the ground. The ground effect may extend as high as a full wingspan's length above the ground.
    a. *Ground effect* is a cushioning effect caused by compression of the air between the ground and the bottom of the wing in general and disruption of the wingtip vortices in particular.
    b. Less thrust is required in **ground effect** than is required out of ground effect.

Chapter 2: Introduction to Airplanes and Aerodynamics 17

2. Ground effect may cause an airplane to float on landings, or permit it to become airborne with insufficient airspeed to stay in flight above the area of ground effect.

   a. An airplane may settle back to the surface abruptly after flying through the ground effect if the pilot has not attained recommended takeoff airspeed.

3. Ground effect reduces both the induced angle of attack and the induced drag on the aircraft.

   a. Wingtip vortices, downwash and upwash, are also reduced.

4. Indicated airspeed is less in ground effect due to a change in pressure around the static source.

AIRPLANE TURN (1 question)

1. The **horizontal component of lift** makes an airplane turn.

   a. To attain this horizontal component of lift, the pilot coordinates rudder, aileron, and elevator.

2. The rudder on an airplane controls the yaw, i.e., rotation about the vertical axis, but **does not** cause the airplane to turn.

AIRPLANE STABILITY (5 questions)

1. An inherently stable airplane **returns to its original condition** (position or attitude) after being disturbed.

   a. It requires less effort to control.

2. The location of the **center of gravity** (CG) with respect to the **center of lift** determines the **longitudinal stability** of an airplane.

3. Airplanes normally pitch down when power is reduced because the resulting lower airspeed and reduced propeller-induced airflow provides less negative lift on the tail. This allows the nose to drop.

4. When the CG in an airplane is located **rear of the CG limit**, the airplane

   a. Develops an inability to recover from stall conditions, and
   b. Becomes less stable and less controllable as airspeed decreases.

TORQUE AND P-FACTOR (3 questions)

1. The **torque effect** (left-turning tendency) is **greatest at low airspeed and high power**, e.g., on takeoff.

2. **P-factor** (asymmetric propeller loading) **causes the airplane to yaw to the left** when at high angles of attack because the right side of the propeller (as seen from the rear) has a higher angle of attack (than the left lower side) and provides more thrust.

Chapter 2: Introduction to Airplanes and Aerodynamics

LOAD FACTOR (7 questions)

1. *Load factor* refers to the additional weight carried by the wings due to centrifugal force.
   a. The amount of excess load that can be imposed on an airplane's wings varies directly with the airplane's speed and the excess lift available.
      1) At low speeds, very little excess lift is available, so very little excess load can be imposed.
      2) At high speeds, the wings' lifting capacity is so great that the load factor can quickly exceed safety limits.
   b. An increased load factor will result in an airplane **stalling at a higher airspeed**.
   c. As bank angle increases, the load factor increases. The wings not only have to carry the airplane's weight, but the centrifugal force as well.

2. On the exam, **a load factor chart** is given with the amount of bank on the horizontal axis (along the bottom of the graph), and the load factor on the vertical axis (up the left side of graph).
   a. Compute the load factor by moving up from the stated degree of bank angle until intersecting the load factor curve. Then move across from the point of intersection to the left side of the graph and read the amount of load factor.
   b. Example load factor chart:

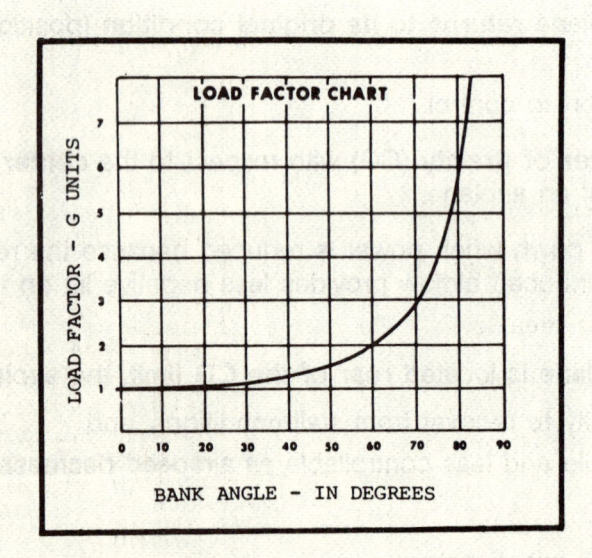

3. Load factor (or G units) is a multiple of the regular weight or, alternatively, a multiple of the force of gravity.
   a. Straight-and-level flight has a load factor at 1.0. (Verify on the chart above.)
   b. A 60° level bank has a load factor of 2.0. Due to centrifugal force, the wings must hold up twice the amount of weight.
   c. A 50° level bank has a load factor of about 1.5.

Chapter 2: Introduction to Airplanes and Aerodynamics

> **QUESTIONS AND ANSWER EXPLANATIONS**
>
> *All of the FAA questions from the written test for the Recreational Pilot certificate relating to the basics of aerodynamics and the material outlined previously are reproduced below in the same modules as the previous outlines. To the immediate right of each question are the correct answer and answer explanation. You should cover these answers and answer explanations while responding to the questions. Refer to the general discussion in Chapter 1 on how to take the examination.*
>
> *Remember that the questions from the FAA's Written Test Book have been reordered by topic, and the topics have been organized into a meaningful sequence. Accordingly, the first line of the answer explanation gives the FAA question number and the citation of the authoritative source for the answer.*

FLAPS AND RUDDER

1.
0115. One of the main purposes in using flaps during the approach and landing is to

A— increase the angle of descent without increasing airspeed.
B— decrease the angle of descent without increasing the airspeed.
C— decrease lift, thus enabling a steeper-than-normal approach to be made.
D— permit a touchdown at a higher indicated airspeed.

Answer (A) is correct (0115). *(PHAK Chap II)*

The practical effect of flaps is to permit a steeper angle of descent without an increase in airspeed.

Answer (B) is incorrect because flaps can increase the angle of descent. Answer (C) is incorrect because flaps increase lift. Answer (D) is incorrect because flaps permit touchdown with no increase in airspeed.

2.
0125. The purpose of wing flaps is to

A— enable the pilot to make steeper approaches to a landing without increasing airspeed.
B— enable the pilot to reduce the speed for the approach to a landing.
C— enlarge or control the wing area to vary the lift.
D— create more drag to utilize power on the approach.

Answer (A) is correct (0125). *(PHAK Chap II)*

Extending the flaps increases the wing camber and increases lift at the same angle of attack. This increases wing lift, enabling the pilot to make steeper approaches to a landing without an increase in airspeed.

Answer (B) is incorrect because the objective is to steepen the flight path, not decrease approach speed. Answer (C) is incorrect because wing area usually remains the same after flap deployment (except for fowler flaps). Answer (D) is incorrect because the increased drag is to steepen the flight path, not utilize power.

3.
0124. The purpose of the rudder on an airplane is to

A— control the yaw.
B— control the overbanking tendency.
C— maintain a crab angle to control drift.
D— maintain the turn after the airplane is banked.

Answer (A) is correct (0124). *(PHAK Chap I)*

The rudder is used to control the yaw about the airplane's vertical axis.

Answers (B) and (D) are incorrect because the aileron controls the bank and maintains the turn after the airplane is banked. The rudder is used only to coordinate the turn. Answer (C) is incorrect because crab angle is maintained by both the ailerons and the rudder, not the rudder alone.

AERODYNAMIC FORCES

4.
0112. When are the four aerodynamic forces that act on an airplane in equilibrium?

A— When the aircraft is at rest on the ground.
B— When the aircraft is accelerating.
C— While the aircraft is decelerating.
D— During unaccelerated flight.

Answer (D) is correct (0112). *(PHAK Chap I)*

During unaccelerated straight-and-level flight, the four aerodynamic forces are in equilibrium.

Answer (A) is incorrect because, when the airplane is at rest on the ground, there are no aerodynamic forces acting on it other than gravity. Answer (B) is incorrect because thrust exceeds drag when the airplane is accelerating. Answer (C) is incorrect because drag exceeds thrust when the airplane is decelerating.

5.
0113. What is the relationship of lift, drag, thrust, and weight when the airplane is in straight-and-level flight?

A— Lift equals drag and thrust equals weight.
B— Lift, drag, and weight equal thrust.
C— Lift and weight equal thrust and drag.
D— Lift equals weight and thrust equals drag.

Answer (D) is correct (0113). *(PHAK Chap I)*

When the airplane is in unaccelerated straight-and-level flight, lift equals weight and thrust equals drag.

Answers (A), (B), and (C) are incorrect because thrust and drag are opposites, and lift and weight are opposites.

ANGLE OF ATTACK

6.
0117. The term angle of attack is defined as the

A— angle between the wing chord line and the relative wind.
B— angle between the airplane's climb angle and the horizon.
C— angle formed by the longitudinal axis of the airplane and the chord line of the wing.
D— specific angle at which the ratio between lift and drag is the highest.

Answer (A) is correct (0117). *(FTH Chap 17)*

The angle of attack is the angle between the wing chord line and the direction of the relative wind. The wing chord line is a straight line from the leading edge to the trailing edge of the wing. The relative wind is the direction of airflow relative to the wing moving through the air.

Answer (B) is incorrect because it does not describe any term. Answer (C) is incorrect because it describes the angle of incidence. Answer (D) is incorrect because it defines the best glide speed and provides the maximum range, i.e., best endurance for a given fuel quantity.

7.
0144. Angle of attack is defined as the angle between the chord line of an airfoil and the

A— horizon.
B— pitch angle of an airfoil.
C— rotor plane of rotation.
D— relative wind.

Answer (D) is correct (0144). *(FTH Chap 17)*

The angle of attack is the angle between the chord line of the airfoil and the relative wind.

Answer (A) is incorrect because the horizon relates to the pitch angle of the aircraft. Answer (B) is incorrect because pitch is used in conjunction with the aircraft or longitudinal axis, not the chord line of the airfoil. If anything, the pitch angle of the airfoil would refer to the angle of incidence. Answer (C) is incorrect because it is a nonsense statement in this context.

Chapter 2: Introduction to Airplanes and Aerodynamics

8.
0116. (Refer to figure 11.) The acute angle A is the angle of

A— dihedral.
B— attack.
C— camber.
D— incidence.

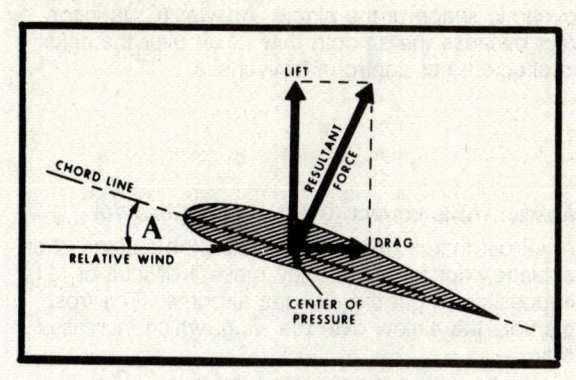

FIGURE 11.—Lift Vector.

Answer (B) is correct (0116). *(FTH Chap 17)*
The angle between the relative wind and the wing chord line is the angle of attack. The wing chord line is a straight line from the leading edge to the trailing edge of the wing.

Answer (A) is incorrect because the dihedral is the angle at which the wings are slanted upward from the root to the tip. Answer (C) is incorrect because camber is the curvature of the airfoil from the leading edge to the trailing edge. The upper camber is on the top of the wing and the lower camber is on the bottom of the wing. Answer (D) is incorrect because the angle of incidence is the acute angle formed by the chord line of the wing and the longitudinal axis of the airplane.

9.
0137. The angle of attack at which an airplane wing stalls will

A— increase if the CG is moved forward.
B— decrease if the CG is moved aft.
C— change with an increase in gross weight.
D— remain the same regardless of gross weight.

Answer (D) is correct (0137). *(FTH Chap 17)*
A given airplane will always stall at the same angle of attack regardless of airspeed, weight, load factor, or density altitude. Each airplane has a particular angle of attack at which the airflow separates from the upper surface of the wing and the stall occurs.

Answers (A), (B), and (C) are incorrect because there is no change in the angle of attack.

STALLS AND SPINS

10.
0138. As altitude increases, the indicated airspeed at which a given airplane stalls in a particular configuration will

A— increase because the air density becomes less.
B— decrease as the true airspeed increases.
C— decrease as the true airspeed decreases.
D— remain the same as at low altitude.

Answer (D) is correct (0138). *(FTH Chap 17)*
The lift, drag, etc. of an airplane are dependent upon air density. As altitude increases, air density decreases, but so does the indicated airspeed. Indicated airspeed is directly related to air density. Accordingly, an airplane will stall in a particular configuration at the same indicated airspeed regardless of altitude.

11.
0139. In what flight condition must an airplane be placed in order to spin?

A— Partially stalled with one wing low and the throttle closed.
B— Placed in a steep diving spiral.
C— Placed in a steep nose-high pitch attitude.
D— Stalled.

Answer (D) is correct (0139). *(AFNA 4)*
In order to enter a spin, an airplane must always first be stalled. Thereafter, the spin is caused when one wing is less stalled than the other wing.

Answer (A) is incorrect because the aircraft must first be fully stalled. Answers (B) and (C) are incorrect because an airplane can be in slow, level flight stalled with cross control and enter a spin.

FROST

12.
0452. Why is frost considered hazardous to flight operation?

A— The increased weight requires a greater takeoff distance.
B— Frost changes the basic aerodynamic shape of the airfoil.
C— Frost decreases control effectiveness.
D— Frost causes early airflow separation resulting in a loss of lift.

Answer (D) is correct (0452). *(AvW Chap 10)*
Frost creates friction, which interferes with the smooth flow of air over the wing surfaces, increasing drag and decreasing lift.

Answer (A) is incorrect because frost is usually thin and light in weight. Answer (B) is incorrect because frost is thin and does not change the aerodynamic shape of the airfoil. Answer (C) is incorrect because the smooth flow of air over the airfoil is affected, not control effectiveness.

13.
0454. Frost which has not been removed from the lifting surfaces of an airplane before flight

A— may prevent the airplane from becoming airborne.
B— will change the camber (curvature of the wing) thereby increasing lift during the takeoff.
C— may cause the airplane to become airborne with a lower angle of attack and at a lower indicated airspeed.
D— would present no problems since frost will blow off when the airplane starts moving during takeoff.

Answer (A) is correct (0454). *(AvW Chap 10)*
Frost that is not removed from the surface of an airplane prior to takeoff may make it difficult or impossible to get the airplane airborne. The frost disrupts the airflow over the wing, which increases drag.

Answer (B) is incorrect because the smoothness of the wing, not its curvature, is affected and lift is decreased. Answer (C) is incorrect because frost requires a higher, not lower, indicated airspeed for takeoff. Answer (D) is incorrect because frost does not blow off. It is frozen onto the wing's surface.

14.
0134. Frost which has not been removed from the wings of an airplane before flight

A— may cause the airplane to become airborne with a lower angle of attack and at a lower indicated airspeed.
B— may make flight difficult or impossible.
C— would present no problems since frost will blow off when the airplane starts moving during takeoff.
D— will change the camber (curvature of the wing) thereby increasing lift during takeoff.

Answer (B) is correct (0134). *(AvW Chap 10)*
Frost that is not removed from the surface of an airplane prior to takeoff may make it difficult or impossible to get the airplane airborne. The frost disrupts the airflow over the wing, which increases drag.

Answer (A) is incorrect because frost requires a higher, not lower, indicated airspeed for takeoff. Answer (C) is incorrect because frost does not blow off. It is frozen onto the wing's surface. Answer (D) is incorrect because the smoothness of the wing, not its curvature, is affected and lift is decreased.

Chapter 2: Introduction to Airplanes and Aerodynamics

GROUND EFFECT

15.
0140. The phenomenon of ground effect is most likely to result in which problem in an airplane?

A— Settling back to the surface abruptly immediately after becoming airborne.
B— Becoming airborne before reaching recommended takeoff speed.
C— Inability to get airborne even though airspeed is sufficient for normal takeoff needs.
D— A rapid rate of sink and absence of normal cushioning during landings.

Answer (B) is correct (0140). *(FTH Chap 17)*

Ground effect may be experienced on takeoff or landing when the airplane is less than its wingspan distance above the surface. A cushioning effect is caused by compression of the air between the bottom of the wing and the ground, giving a little extra lift when flying near the surface. Ground effect may allow a pilot to pull the airplane from the ground before it has reached flying speed. It may then settle back to the surface abruptly after flying through the ground effect area if the pilot does not allow the airplane to develop some speed before attempting a climb.

Answer (A) is incorrect because the airplane will settle back to the surface after flying through the ground effect, not immediately after becoming airborne. Answer (C) is incorrect because ground effect would not hamper the airplane from becoming airborne if the airspeed were sufficient for normal takeoff. Ground effect helps the airplane become airborne before reaching the recommended takeoff speed. Answer (D) is incorrect because ground effect would most likely cause a floating condition or a cushioning effect on landing.

16.
0141. Which adverse effect must a pilot be aware of as a result of the phenomenon of ground effect during takeoff?

A— Difficulty in getting airborne even though airspeed is sufficient for normal takeoff.
B— Becoming airborne before reaching recommended takeoff speed.
C— Settling back to the surface abruptly immediately after becoming airborne.
D— Difficulty in climbing the first 20 feet after takeoff.

Answer (B) is correct (0141). *(FTH Chap 17)*

Ground effect provides a cushioning effect at altitudes of less than the wingspan, which will allow an airplane to become airborne prematurely. It will then settle back to the surface abruptly after flying through the ground effect if the pilot does not allow the airplane to develop sufficient airspeed before attempting a climb.

Answer (A) is incorrect because ground effect allows an airplane to become airborne at below takeoff speed. Answer (C) is incorrect because the airplane will settle back to the surface after flying through the ground effect, not immediately after becoming airborne. Answer (D) is incorrect because ground effect makes climbing the first 20 feet after takeoff easier.

17.
0143. An airplane is usually affected by ground effect at what height above the surface?

A— Between 100 and 200 feet above the surface in calm wind conditions.
B— Less than half of the airplane's wingspan above the surface.
C— Twice the length of the airplane's wingspan above the surface.
D— Three or four times the airplane's wingspan.

Answer (B) is correct (0143). *(FTH Chap 17)*

Ground effect is said to exist roughly between the surface and about one-half of the wingspan from the surface.

Answers (A), (C), and (D) are incorrect because ground effect does not extend to twice or more of the airplane's wingspan above the surface.

Chapter 2: Introduction to Airplanes and Aerodynamics

18.
0142. The effect of floating caused by the phenomenon of ground effect will be most realized during an approach to land when at

A— less than the length of the airplane's wingspan above the surface.
B— twice the length of the airplane's wingspan above the surface.
C— a higher-than-normal angle of attack.
D— speeds approaching a stall.

Answer (A) is correct (0142). *(FTH Chap 17)*

Ground effect is most pervasive when the airplane is within one-half of the length of the airplane's wingspan above the surface. It may extend as high as a full wingspan length above the surface. In this area, a cushioning effect is caused by compression of the air between the bottom of the wing and the ground.

Answer (B) is incorrect because ground effect generally only extends up to one wingspan length. Answers (C) and (D) are incorrect because ground effect is dependent on height above the ground, not angle of attack or airspeed.

AIRPLANE TURN

19.
0114. What makes an airplane turn?

A— Centrifugal force.
B— Rudder and aileron.
C— Horizontal component of lift.
D— Rudder, aileron, and elevator.

Answer (C) is correct (0114). *(PHAK Chap I)*

When the wings of an airplane are not level, the lift is not vertical and tends to pull the airplane toward the direction of the lower wing. An airplane is turned when the pilot coordinates rudder, aileron, and elevator to bank in order to attain a horizontal component of lift.

Answer (A) is incorrect because the horizontal component of lift counteracts centrifugal force, which would have the plane fly in a straight line. Answers (B) and (D) are incorrect because it is the aerodynamic force of lift that turns the plane, although the rudder, aileron, and elevator are used to accomplish this.

AIRPLANE STABILITY

20.
0129. An airplane said to be inherently stable will

A— be difficult to stall.
B— require less effort to control.
C— not spin.
D— not overbank during steep turns.

Answer (B) is correct (0129). *(PHAK Chap I)*

An inherently stable airplane will usually return to the original condition of flight (except when in a bank) if disturbed by a force such as turbulence of air. Thus, an inherently stable airplane will require less effort to control than an inherently unstable one.

Answers (A) and (C) are incorrect because inherent stability does not preclude spinning or stalling. Answer (D) is incorrect because, in most airplanes, the lift differential overbalances the lateral stability, and counteractive pressure on the ailerons is necessary to keep the bank from steepening.

Chapter 2: Introduction to Airplanes and Aerodynamics

21.
0130. What determines the longitudinal stability of an airplane?

A— The location of the CG with respect to the center of lift.
B— The effectiveness of the horizontal stabilizer, rudder, and rudder trim tab.
C— The relationship of thrust and lift to weight and drag.
D— The dihedral, angle of sweepback, and the keel effect.

Answer (A) is correct (0130). *(PHAK Chap I)*
The location of the center of gravity with respect to the center of lift determines, to a great extent, the longitudinal stability of the airplane. Positive stability is attained by having the center of lift behind the center of gravity. Then the tail provides negative lift, creating a downward tail force, which counteracts the nose's tendency to pitch down.
Answer (B) is incorrect because the rudder and rudder trim tab control the yaw, not the pitch. Answer (C) is incorrect because the relationship of thrust and lift to weight and drag affects speed and altitude. Answer (D) is incorrect because the dihedral, angle of sweepback, and keel effect all affect lateral stability.

22.
0131. What causes an airplane (except a T-tail) to pitch nosedown when power is reduced and controls are not adjusted?

A— The CG shifts forward when thrust and drag are reduced.
B— The downwash on the elevators from the propeller slipstream is reduced and elevator effectiveness is reduced.
C— When thrust is reduced to less than weight, lift is also reduced and the wings can no longer support the weight.
D— The upwash on the wings from the propeller slipstream is reduced and angle of attack is reduced.

Answer (B) is correct (0131). *(PHAK Chap I)*
When power is reduced, both the airspeed and propeller-induced airflow diminish on the horizontal stabilizer (i.e., less negative lift on the tail), and the nose pitches downward.
Answer (A) is incorrect because the CG depends upon weight within the plane, not the amount of thrust or drag. Answer (C) is incorrect because thrust is the opposite of drag, not weight. Answer (D) is incorrect because the angle of attack is the relationship of the wing to the relative air, not to the propeller slipstream.

23.
0132. An airplane has been loaded in such a manner that the CG is located aft of the CG limit. One undesirable flight characteristic a pilot might experience with this airplane would be

A— a longer takeoff run.
B— the inability to recover from a stalled condition.
C— stalling at higher-than-normal airspeed.
D— the inability to flare during landings.

Answer (B) is correct (0132). *(FTH Chap 17)*
The recovery from a stall in any airplane becomes progressively more difficult as its center of gravity moves backward. Generally, airplanes become less controllable, especially at slow flight speeds, as the center of gravity is moved backward.
Answers (A) and (C) are incorrect because an airplane with an aft CG limit has less drag; i.e., less force must be put on the tail. Also, the aft CG results in a lower stalling speed. Answer (D) is incorrect because it is easier to flare during landings with weight in the back.

24.
0133. Loading an airplane to the most aft CG causes the airplane to be

A— less controllable at slow speeds than at high speeds.
B— less controllable at high speeds than at low speeds.
C— less controllable at moderate to high airspeeds than at low airspeeds.
D— more controllable at all airspeeds.

Answer (A) is correct (0133). *(FTH Chap 17)*
An airplane generally becomes less controllable, especially at slow airspeeds, as the C.G. is moved aft. There is an aft C.G. limit due to control and stability problems, and there is less control with less air (i.e., at slow airspeeds) moving over the elevator.
Answers (B) and (C) are incorrect because adverse controllability occurs more at slower than at higher airspeeds. Answer (D) is incorrect because the controllability decreases, not increases.

TORQUE AND P-FACTOR

25.
0135. P-factor or asymmetric propeller loading causes the airplane to

A— be unstable around the lateral axis.
B— be unstable around the vertical and lateral axes.
C— yaw to the left when at high angles of attack.
D— yaw to the left when at high speeds.

Answer (C) is correct (0135). *(PHAK Chap I)*

P-factor or asymmetric propeller loading occurs when an airplane is flown at a high angle of attack because the downward-moving blade on the right side of the propeller (as seen from the rear) has a higher angle of attack, which creates higher thrust than the upward moving blade on the left. Thus, the airplane yaws around the vertical axis to the left.

Answers (A) and (B) are incorrect because the P-factor does not cause instability. Answer (D) is incorrect because, at high speeds, an airplane is not at a high angle of attack.

26.
0123. In what airspeed and power condition is torque effect the greatest in a single-engine airplane?

A— High airspeed, high power.
B— High airspeed, low power.
C— Low airspeed, high power.
D— Low airspeed, low power.

Answer (C) is correct (0123). *(PHAK Chap I)*
NOTE: FAA Question 0136 is an exact duplicate.

The effect of torque increases in direct proportion to engine power and inversely to airspeed. Thus, at low airspeeds and high power settings, torque is the greatest. Torque is a combination of forces that causes a twisting or rotating motion of the airplane to the left (looking toward the front of the airplane from the rear).

27.
0122. The left turning tendency of an airplane caused by P-factor is the result of the

A— clockwise rotation of the engine and the propeller turning the airplane counterclockwise.
B— propeller blade descending on the right, producing more thrust than the ascending blade on the left.
C— gyroscopic forces applied to the rotating propeller blades acting 90° in advance of the point the force was applied.
D— spiral characteristics of the slipstream air being forced rearward by the rotating propeller.

Answer (B) is correct (0122). *(PHAK Chap I)*

Asymmetric propeller loading (P-factor) occurs when the airplane is flown at a high angle of attack because the downward-moving blade on the right side of the propeller (as seen from the rear) has a higher angle of attack, which creates higher thrust than the upward moving blade on the left. Thus, the airplane yaws around the vertical axis to the left.

Answer (A) is a description of reactive force. Answer (C) is a description of gyroscopic precession. Answer (D) is a description of spiraling slipstream. The four alternative answers are the four forces resulting in torque (twisting and turning of the airplane to the pilot's left).

LOAD FACTOR

28.
0118. (Refer to figure 12.) If an airplane weighs 2,300 pounds, what approximate weight would the airplane structure be required to support during a 60° banked turn while maintaining altitude?

A— 2,300 pounds.
B— 3,400 pounds.
C— 4,600 pounds.
D— 5,200 pounds.

Answer (C) is correct (0118). *(PHAK Chap I)*

Note on the chart that, at a 60° bank angle, the load factor is 2. Thus, a 2,300 lb plane would exert 4,600 lb of force (2,300 x 2) on the wings in a 60° bank turn.

Chapter 2: Introduction to Airplanes and Aerodynamics

LOAD FACTOR CHART

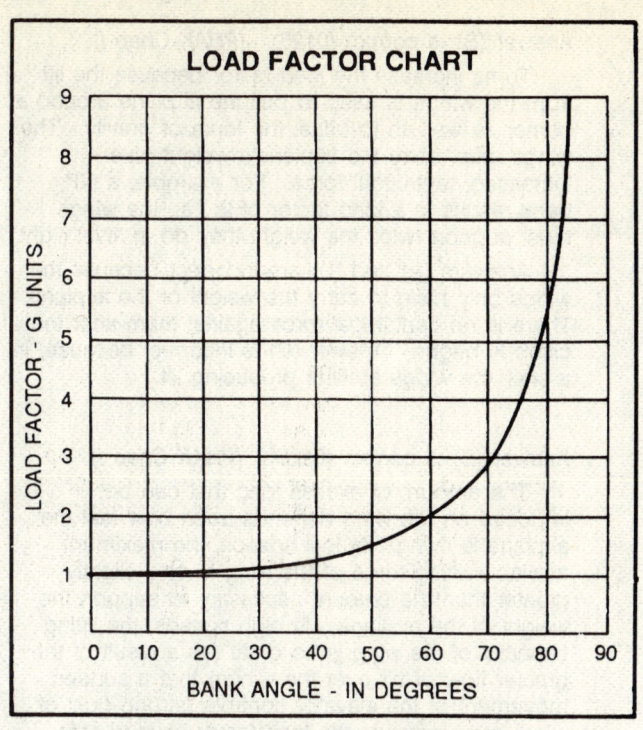

FIGURE 12.—Load Factor Chart.

29.
0119. (Refer to figure 12.) If an airplane weighs 3,300 pounds, what approximate weight would the airplane structure be required to support during a 30° banked turn while maintaining altitude?

A— 1,200 pounds.
B— 3,100 pounds.
C— 3,960 pounds.
D— 7,220 pounds.

Answer (C) is correct (0119). *(PHAK Chap I)*
Look on the chart to see that, at a 30° bank angle, the load factor is about 1.2. Thus, a 3,300 lb airplane in a 30° bank would require its wings to support 3,960 lb (3,300 x 1.2).

30.
0120. (Refer to figure 12.) If an airplane weighs 5,400 pounds, what approximate weight would the airplane structure be required to support during a 55° banked turn while maintaining altitude?

A— 5,400 pounds
B— 6,720 pounds.
C— 9,180 pounds.
D— 10,800 pounds.

Answer (C) is correct (0120). *(PHAK Chap I)*
Look on the chart under 55° and note that the load factor curve is less than 2 but more than 1.5. Of the answer choices, 9,180 lb (5,400 x 1.7) is the only possible correct answer.

31.
0121. (Refer to figure 12.) The maximum bank that could be made during a level turn without exceeding the maximum positive load factor of a utility category airplane (+4.4 G units) is

A— 71°.
B— 73°.
C— 77°.
D— 83°.

Answer (C) is correct (0121). *(PHAK Chap I)*
Utility category airplanes are structured to carry 4.4 Gs, as given in the question. On the load factor chart, 4.4 along the vertical axis is more than 75° of bank but less than 80°. Thus, the correct answer is 77°.

32.
0128. Which basic flight maneuver increases the load factor on an airplane as compared to straight-and-level flight?

A— Climbs.
B— Turns.
C— Stalls.
D— Slips.

Answer (B) is correct (0128). *(PHAK Chap I)*

Turns increase the load factor because the lift from the wings is used to pull the airplane around a corner as well as to offset the force of gravity. The wings must carry the airplane's weight plus increased centrifugal force. For example, a 60° bank results in a load factor of 2; i.e., the wings must support twice the weight they do in level flight.

Answers (A) and (D) are incorrect because the wings only have to carry the weight of the airplane. There is no centrifugal force against them after the climb is begun. Answer (C) is incorrect because, in a stall, the wings are not producing lift.

33.
0126. The amount of excess load that can be imposed on the wing of an airplane depends upon the

A— position of the CG.
B— speed of the airplane.
C— abruptness at which the load is applied.
D— angle of attack at which the airplane will stall.

Answer (B) is correct (0126). *(PHAK Chap I)*

The amount of excess load that can be imposed on the wing depends upon how fast the airplane is flying. At low speeds, the maximum available lifting force of the wing is only slightly greater than the amount necessary to support the weight of the airplane. At high speeds, the lifting capacity of the wing is so great (as a result of the greater flow of air over the wings) that a sudden movement of the elevator controls (strong gust of wind) may increase the load factor beyond safe limits. This is why maximum speeds are established by airplane manufacturers.

Answer (A) is incorrect because the position of the CG affects the stability of the airplane but not the total load the wings can support. Answer (C) is incorrect because it is the amount of load, not the abruptness of the load, that is limited. However, the abruptness of the maneuver can affect the amount of the load. Answer (D) is incorrect because the angle of attack at which an airplane stalls remains constant regardless of load.

34.
0127. During an approach to a stall, an increased load factor will cause the airplane to

A— stall at a higher airspeed.
B— have a tendency to spin.
C— be more difficult to control.
D— have a tendency to yaw and roll as the stall is encountered.

Answer (A) is correct (0127). *(PHAK Chap I)*

The greater the load (whether from gross weight and gravity or from centrifugal force), the more lift is required. Therefore, an airplane will stall at higher airspeeds when the load factor is increased.

Answers (B), (C), and (D) are incorrect because the load factor does not directly affect spins, control, or yaw and roll.

CHAPTER THREE
AIRPLANE PERFORMANCE

Density Altitude	*(7 questions)*	29, 34
Density Altitude Computations	*(7 questions)*	30, 36
Takeoff Distance	*(4 questions)*	31, 38
Cruise Performance	*(5 questions)*	32, 40
Crosswind Components	*(6 questions)*	32, 42
Landing Distance	*(10 questions)*	33, 44
Wingtip Vortices	*(4 questions)*	33, 47

This chapter contains outlines of major concepts tested, all FAA test questions and answers regarding airplane performance, and an explanation of each answer. The subtopics or modules within this chapter are listed above, followed in parentheses by the number of questions from the FAA written test pertaining to that particular module. The two numbers following the parentheses are the page numbers on which the outline and questions begin for that module.

CAUTION: Recall that the sole purpose of this book is to expedite your passing the FAA written test for the recreational pilot certificate. Accordingly, all extraneous material (i.e., topics or regulations not directly tested on the FAA written test) is omitted, even though much more information and knowledge are necessary to fly safely. This additional material is presented in *RECREATIONAL PILOT FLIGHT MANEUVERS* and *PRIVATE PILOT HANDBOOK*, available from Aviation Publications, Inc. See pages 208 to 210 for more information and an order form.

Many of the topics in this chapter require interpretation of graphs and charts. Graphs and charts pictorially describe the relationship between two or more variables. Thus, they are a substitute for solving one or more equations. Each time you must interpret (i.e., get an answer from) a graph or chart, you should

1. Understand clearly what is required, e.g., landing roll distance, etc.
2. Analyze the chart or graph to determine the variables involved, including
    a. Labeled sides (axes) of the graph or chart.
    b. Labeled lines within the graph or chart.
3. Plug the data given in the question into the graph or chart.
4. Finally, determine the value of the item required in the question.

DENSITY ALTITUDE (7 questions)

1. *Density altitude* is a measurement of the density of the air in terms of altitude.
    a. **Air density varies inversely with altitude**; i.e., air is very dense at low altitudes and less dense at high altitudes.

Chapter 3: Airplane Performance

- b. Temperature, humidity, and barometric pressure also affect air density.
    1) The **scale of air density to altitude** was made using a constant temperature, humidity, and barometric pressure. These are referred to as standard temperature, humidity, and barometric pressure.
        a) Standard temperature at sea level is 15°C (59°F).
        b) Standard pressure at sea level is 29.92 inches Hg.
    2) When these factors are not at standard (which is almost always), density altitude will not be the same as true altitude.

2. You are also required to know how barometric pressure, temperature, and humidity affect density altitude. Visualize the following:
    a. **As barometric pressure increases,** the air becomes more compressed and compact. This is an increase in density. Air density is higher if the pressure is high, which means a lower density altitude.
        1) Density altitude is increased by a decrease in pressure.
        2) Density altitude is decreased by an increase in pressure.
    b. **As temperature increases,** the air molecules become more active, bouncing against each other, and taking up more room. This is a decrease in density, which means a higher density altitude. Remember, air is normally less dense at higher altitudes.
        1) Density altitude is increased by an increase in temperature.
        2) Density altitude is decreased by a decrease in temperature.
    c. **As relative humidity increases,** the air molecules are affected the same as by temperature increases. A decrease in density occurs, resulting in a higher density altitude.
        1) Density altitude is increased by an increase in humidity.
        2) Density altitude is decreased by a decrease in humidity.

3. Said another way, density altitude varies **directly with temperature and humidity,** and **inversely with barometric pressure**:
    a. Cold, dry air and higher barometric pressure = low density altitude.
    b. Hot, humid air and lower barometric pressure = high density altitude.

4. **Pressure altitude** is based on standard temperature. Therefore, density altitude will exceed pressure altitude if the temperature is above standard.

5. The **primary reason for computing density altitude** is to determine airplane performance.
    a. High density altitude reduces an airplane's performance.
    b. For example, climb performance is less and takeoff distance is longer.
    c. Propellers also have less efficiency at high density altitude because there is less air for the propeller to pull against.
    d. However, the same indicated airspeed is used for takeoffs and landings regardless of altitude or air density because the airspeed indicator is also directly affected by air density.

DENSITY ALTITUDE COMPUTATIONS (7 questions)

1. Density altitude is determined most easily by **first finding the pressure altitude** (altitude if your altimeter is set to 29.92) and adjusting for the temperature.
    a. The adjustment is made using your flight computer or a density altitude chart. This part of the FAA test requires you to use such a chart.

Chapter 3: Airplane Performance

    b. The **second step** is to **adjust pressure altitude for nonstandard temperature.**

2. When using a density altitude chart (see Figure 1 on page 36),

    a. Adjust the airport elevation to pressure altitude based upon the actual altimeter setting in relation to the standard altimeter setting of 29.92.

        1) On the chart, the correction in feet is provided for different altimeter settings.

    b. To adjust the pressure altitude for nonstandard temperature, plot the intersection of the actual air temperature (listed on the horizontal axis of the chart) with the pressure altitude lines that slope upward and to the right. The vertical coordinate (listed on the vertical axis of the chart) of the intersection is the density altitude.

    c. EXAMPLE. Outside air temperature .... 90°F
                   Altimeter setting .... 30.20 inches Hg
                   Airport elevation .... 4,725 feet

Referring to Figure 1 on page 36, you determine the density altitude to be approximately 7,400 feet. This is found as follows:

        1) The altimeter setting of 30.20 requires a -257 altitude correction factor.

        2) Subtract 257 from field elevation of 4,725 ft to obtain pressure altitude of 4,468 ft.

        3) Locate 90°F on the bottom axis of the chart and move up to intersect the diagonal pressure altitude line of 4,468 ft.

        4) Move horizontally to the left axis of the chart to obtain the density altitude of about 7,400 ft.

        5) Note that while true altitude is 4,725 ft, density altitude is about 7,400 ft!

TAKEOFF DISTANCE (4 questions)

1. **Takeoff distance performance** is displayed in the airplane operating manual either

    a. In chart form or
    b. On a graph

2. If a graph, it is usually presented in terms of density altitude. Thus, one must first adjust for temperature and pressure altitude.

    a. In the graph used on this exam (see Figure 5 on page 39), the first section on the left uses outside air temperature and pressure altitude to obtain density altitude.

        1) The vertical line sloping slightly to the left is the standard temperature line, which you use when the question calls for standard temperature.

    b. The second section of the graph to the right (which contains three slightly curved vertical lines) determines "ground roll."

    c. The third section of the graph, to the right of the "ground roll" lines, determines "distance over 50-ft barrier." It also contains three slightly curved vertical lines.

    d. EXAMPLE. Given an outside air temperature of 80°F, a pressure altitude of 1,500 ft, a takeoff weight of 2,325 lb, and a headwind component of 15 kts, find the ground roll and the total take-off distance over a 50-ft obstacle. Use Figure 5 on page 39.

31

e. The solution to the example problem is marked with the dotted arrows on the graph. Move straight up from 80°F to the pressure altitude of 1,500 ft. Move horizontally to the right to the first (15 kt headwind) "ground roll" line. From this intersection, go down vertically to determine the "ground roll" of 1,600 ft. Continue to the right to the first (15 kt headwind) "over 50-ft barrier" line. From that intersection, proceed down vertically to determine the total takeoff distance over a 50-ft obstacle of 2,100 ft.

CRUISE PERFORMANCE (5 questions)

1. Cruise power settings are tested by use of a cruise performance chart (see Figure 4 on page 40).
    a. It is based on density altitude (first column), and
    b. RPM (second column).
    c. Based on altitude and RPM, the following are predicted:
        1) Percent brake horsepower (BHP)
        2) Fuel flow
        3) True airspeed (TAS)
        4) Range

2. The FAA test questions gauge your ability to **find values on the chart** and **interpolate between lines**.
    a. EXAMPLE. A value for 10,000 ft would be 50% of the distance between the number for 9,500 ft and the number for 10,500 ft.

CROSSWIND COMPONENTS (6 questions)

1. Airplanes have an upper limit to the amount of direct crosswind in which they can land. Crosswinds of less than 90° can be converted into a 90° component by the use of a graph. Variables on the crosswind component graphs are
    a. Angle between wind and runway.
        1) The runway number indicates the magnetic degrees of the runway by adding a zero, e.g., Runway 6 means a 60° heading.
    b. Knots of total wind velocity.

    Both variables are plotted on the graph. The coordinates on the vertical and horizontal axes of the graph will indicate the headwind and crosswind components of a quartering wind.

2. Refer to the crosswind component graph (Figure 2 on page 42).
    a. Note the example on the graph of a 40-kt wind at a 30° angle.
    b. Find the 30° wind angle line. This is the angle between the wind direction and runway direction, e.g., Runway 18 and wind from 210°.
    c. Find the 40-kt wind velocity arc. Note the intersection of the wind arc and the 30° angle line.
        1) Drop straight down to determine the crosswind component of 20 kts; i.e., landing in this situation would be like having a direct crosswind of 20 kts.
        2) Move horizontally to the left to determine the headwind component of 35 kts; i.e., landing in this situation would be like having a headwind of 35 kts.

Chapter 3: Airplane Performance

LANDING DISTANCE (10 questions)

1. **Required landing distances differ** at various altitudes and temperatures due to changes in air density.

    a. However, indicated airspeed for landing is the same at all altitudes.

2. **Landing distance information** is given in airplane operating manuals, in chart or graph form, to adjust for headwind, temperature, and dry grass runways.

3. It is **imperative** that you distinguish between distances for clearing a 50-ft obstacle and no 50-ft obstacle at the beginning of the runway (the latter is described as the ground roll).

4. See Figure 6 on page 45 for an example landing performance graph. It is used in the same manner as the takeoff distance graph (Figure 5) discussed on page 31 and printed on page 39.

5. Refer to Figure 7 on page 47, which is an airplane landing distance table.

    a. It has been computed for landing with no wind, at standard temperature, and at pressure altitude.

    b. The notes at the bottom tell you how to adjust for wind, nonstandard temperature, and a grass runway.

        1) Note 1 says to decrease the distance for a headwind. If there is a tailwind, increase the distance using the same formula.

    c. EXAMPLE. Given standard air temperature, 8 kts headwind, and pressure altitude of 2,500 ft, find both the ground roll and the landing distance to clear a 50-ft obstacle.

        1) On the table (Figure 7) for 2,500 ft, at standard temperature with no wind, the ground roll is 470 ft and the distance to clear a 50-ft obstacle is 1,135 ft. These amounts must be decreased by 20% because of the headwind (8 kts ÷ 4 x 10% = 20%). Therefore, the ground roll is 376 ft (470 x 80%) and the distance to clear a 50-ft obstacle is 908 ft (1,135 x 80%).

WINGTIP VORTICES (4 questions)

1. Wingtip vortices (wake turbulence) **are created** when airplanes develop lift.

2. Wingtip vortices **are most pronounced** when the airplane is at low speeds during climbs or approaches for landings (i.e., heavy and slow).

3. Wingtip vortice turbulence tends to sink into the flightpath of airplanes operating below the airplane generating the turbulence.

    a. Thus, you should **fly above** the flightpath of a large jet rather than below.
    b. You should also **fly upwind** rather than downwind of the flightpath, since crosswinds will drift the vortices.

4. **The most dangerous wind**, when avoiding wake turbulence, is the light quartering tailwind. It will push the vortices into your touchdown zone, even if you are executing proper procedures.

Chapter 3: Airplane Performance

QUESTIONS AND ANSWER EXPLANATIONS

All of the FAA questions from the written test for the Recreational Pilot certificate relating to airplane performance and the material outlined previously are reproduced below in the same modules as the previous outlines. To the immediate right of each question are the correct answer and answer explanation. You should cover these answers and answer explanations while responding to the questions. Refer to the general discussion in Chapter 1 on how to take the examination.

Remember that the questions from the FAA's Written Test Book have been reordered by topic, and the topics have been organized into a meaningful sequence. Accordingly, the first line of the answer explanation gives the FAA question number and the citation of the authoritative source for the answer.

DENSITY ALTITUDE

1.
0401. What are the standard temperature and pressure values for sea level?

A— 15 °C and 29.92" Hg.
B— 15 °C and 1013.2" Hg.
C— 59 °F and 29.92 millibars.
D— 59 °C and 1013.2 millibars.

Answer (A) is correct (0401). *(AvW Chap 3)*

The standard temperature and pressure values for sea level are 15°C and 29.92" Hg. This is equivalent to 59°F and 1013.2 millibars of mercury.

2.
0008. What is the primary reason for computing density altitude?

A— To determine pressure altitude.
B— To determine aircraft performance.
C— To establish flight levels (FL's) above 18,000 feet MSL.
D— To ensure safe cruising altitude over mountainous terrain.

Answer (B) is correct (0008). *(AvW Chap 3)*

Density altitude concerns the denseness of the air surrounding the airplane. As barometric pressure increases, air becomes more dense. As air becomes warmer, it becomes less dense. Density altitude is the altitude in the standard atmosphere at which air density is the same as where you are. Density altitude is not an altitude, per se. Rather it is an index to aircraft performance. Low density altitude increases performance; high density altitude decreases performance.

Answer (A) is incorrect because density altitude is computed using pressure altitude. Answer (C) is incorrect because flight levels are based on pressure altitude. Answer (D) is incorrect because indicated (or true) altitude, not density altitude, is used in determining safe flight altitudes over mountains.

3.
0007. Which factor would tend to increase the density altitude at a given airport?

A— Increasing barometric pressure.
B— Increasing ambient temperature.
C— Decreasing ambient temperature.
D— Decreasing relative humidity.

Answer (B) is correct (0007). *(AvW Chap 3)*

When air temperature is higher than standard, density altitude increases. At a higher temperature, the air is less dense which occurs at higher altitudes in a standard atmosphere.

Answers (A), (C), and (D) are incorrect because density altitude varies inversely with pressure and directly with humidity and temperature.

Chapter 3: Airplane Performance

4.
0017. What effect does high density altitude as compared to low density altitude have on propeller efficiency?

A— Increased efficiency due to less friction on the propeller blades.
B— Reduced efficiency because the propeller exerts less force than at lower density altitudes.
C— Increased efficiency because the propeller exerts more force on the thinner air.
D— Reduced efficiency due to the increased force of the thinner air on the propeller.

Answer (B) is correct (0017). *(AvW Chap 3)*
The propeller produces thrust in proportion to the mass of air being accelerated through the rotating blades. If the air is less dense, the propeller efficiency is decreased. Remember, higher density altitude refers to less dense air.

Answers (A) and (C) are incorrect because there is decreased, not increased, efficiency. The propeller exerts less force on thinner air. Answer (D) is incorrect because the propeller exerts force on the air rather than the relative air exerting force on the propeller.

5.
0006. What effect does high density altitude have on aircraft performance?

A— Increases engine performance.
B— Reduces an aircraft's climb performance.
C— Decreases the runway length required for takeoff.
D— Increases lift because the light air exerts less force on the airfoils.

Answer (B) is correct (0006). *(PHAK Chap V)*
High density altitude reduces all aspects of an airplane's performance, including climb performance.

Answer (A) is incorrect because engine performance is decreased. Answer (C) is incorrect because takeoff runway length is increased. Answer (D) is incorrect because lift is decreased.

6.
0021. Which combination of atmospheric conditions will reduce aircraft takeoff and climb performance?

A— High temperature, low relative humidity, and low density altitude.
B— Low temperature, low relative humidity, and low density altitude.
C— High temperature, high relative humidity, and high density altitude.
D— Low temperature, high relative humidity, and high density altitude.

Answer (C) is correct (0021). *(PHAK Chap V)*
Takeoff and climb performance are reduced by high density altitude. High density altitude is a result of high temperatures and high relative humidity.

Answers (A), (B), and (D) are incorrect because low temperature, low relative humidity, and low density altitude each improve airplane performance.

7.
0020. If the outside air temperature at a given altitude is warmer than standard, the density altitude is

A— higher than true altitude, but lower than pressure altitude.
B— lower than true altitude.
C— higher than the pressure altitude.
D— lower than pressure altitude, but approximately equal to the true altitude.

Answer (C) is correct (0020). *(AvW Chap 3)*
When temperature increases, the air molecules become more active, bouncing against each other and taking up more room. This is a decrease in density (fewer particles in a given space), which means a higher density altitude. Pressure altitude is based on standard temperature, so it does not increase with temperature.

Answers (A) and (D) are incorrect because density altitude varies from pressure altitude directly as temperature varies from standard. Answers (B) and (D) are incorrect because as temperature increases, so does density altitude compared to true altitude.

DENSITY ALTITUDE COMPUTATIONS

8.
0022. (Refer to figure 1.) What is the effect of a temperature increase from 25 °F to 50 °F on the density altitude if the pressure altitude remains at 5,000 feet?

A— 1,000-foot increase.
B— 1,200-foot increase.
C— 1,400-foot increase.
D— 1,650-foot increase.

Answer (D) is correct (0022). *(PHAK Chap IV)*
Increasing the temperature from 25°F to 50°F, given a pressure altitude of 5,000 ft, requires you to find the 5,000-ft line on the density altitude chart at the 25°F level. At this point, the density altitude is approximately 3,850 ft. Then move up the 5,000-ft line to 50°F, where the density altitude is approximately 5,500 ft. There is about a 1,650-ft increase (5,500 - 3,850 ft).

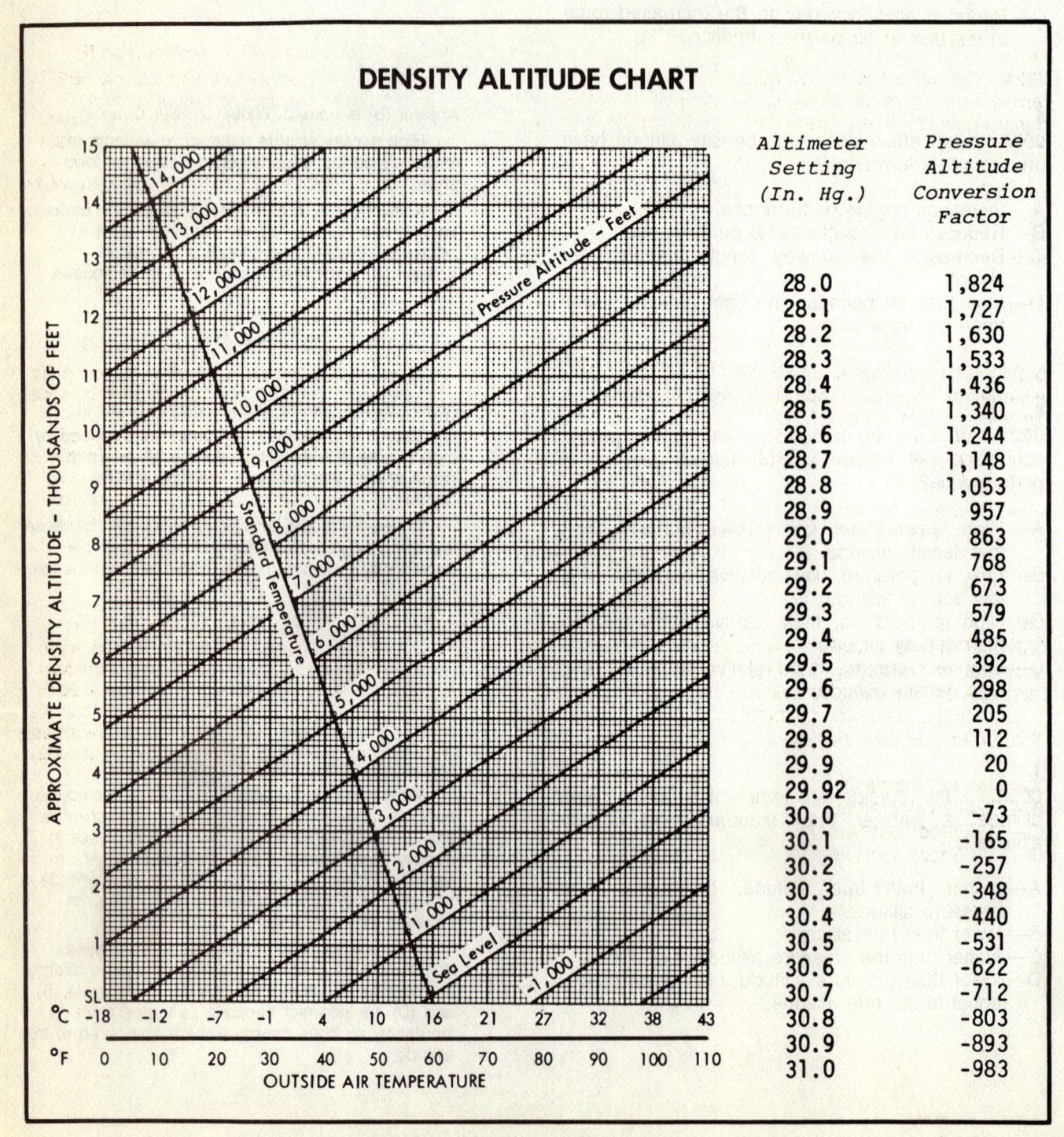

DENSITY ALTITUDE CHART

Altimeter Setting (In. Hg.)	Pressure Altitude Conversion Factor
28.0	1,824
28.1	1,727
28.2	1,630
28.3	1,533
28.4	1,436
28.5	1,340
28.6	1,244
28.7	1,148
28.8	1,053
28.9	957
29.0	863
29.1	768
29.2	673
29.3	579
29.4	485
29.5	392
29.6	298
29.7	205
29.8	112
29.9	20
29.92	0
30.0	-73
30.1	-165
30.2	-257
30.3	-348
30.4	-440
30.5	-531
30.6	-622
30.7	-712
30.8	-803
30.9	-893
31.0	-983

FIGURE 1.—Density Altitude Chart.

Chapter 3: Airplane Performance

9.
0023. (Refer to figure 1.) Determine the pressure altitude at an airport that is 3,563 feet MSL with an altimeter setting of 29.96.

A— 3,507 feet.
B— 3,527 feet.
C— 3,556 feet.
D— 3,639 feet.

Answer (B) is correct (0023). *(PHAK Chap IV)*
Pressure altitude is determined by adjusting the altimeter setting to 29.92" Hg, i.e., adjusting for nonstandard pressure. On paper, this is the true altitude plus or minus the pressure altitude conversion factor (based on current altimeter setting). On the chart, an altimeter setting of 30.0 requires you to subtract 73 ft to determine pressure altitude (note that at 29.92, nothing is subtracted because that is pressure altitude). Since 29.96 is halfway between 29.92 and 30.0, you need only subtract 36 (-73 ÷ 2) from 3,563 ft to obtain a pressure altitude of 3,527 ft (3,563 - 36).

10.
0024. (Refer to figure 1.) What is the effect of a temperature increase from 30 °F to 50 °F on the density altitude if the pressure altitude remains at 3,000 feet?

A— 900-foot increase.
B— 1,100-foot decrease.
C— 1,300-foot increase.
D— 1,500-foot increase.

Answer (C) is correct (0024). *(PHAK Chap IV)*
Increasing the temperature from 30°F to 50°F, given a constant pressure altitude of 3,000 ft, requires you to find the 3,000-ft line on the density altitude chart at the 30°F level. At this point, the density altitude is approximately 1,650 ft. Then move up the 3,000 ft line to 50°F, where the density altitude is approximately 2,950 ft. There is an approximate 1,300-ft increase (2,950 - 1,650 ft).

11.
0025. (Refer to figure 1.) Determine the pressure altitude at an airport that is 1,386 feet MSL with an altimeter setting of 29.97.

A— 1,341 feet.
B— 1,451 feet.
C— 1,562 feet.
D— 1,684 feet.

Answer (A) is correct (0025). *(PHAK Chap IV)*
Pressure altitude is determined by adjusting the altimeter setting to 29.92" Hg. On paper, this is the true altitude plus or minus the pressure altitude conversion factor (based on current altimeter setting). Since 29.97 is not a number given on the conversion chart, you must interpolate. Compute 5/8 of -73 (since 29.97 is 5/8 of the way between 29.92 and 30.0), which is 45. Subtract 45 ft from 1,386 ft to obtain a pressure altitude of 1,341 ft.

12.
0026. (Refer to figure 1.) What is the effect of a temperature decrease and a pressure altitude increase on the density altitude from 90 °F and 1,250 foot pressure altitude to 60 °F and 1,750 foot pressure altitude?

A— 500-foot increase.
B— 500-foot decrease.
C— 1,300-foot increase.
D— 1,300-foot decrease.

Answer (D) is correct (0026). *(PHAK Chap IV)*
The requirement is the effect of a temperature decrease and a pressure altitude increase on density altitude. First, find the density altitude at 90°F and 1,250 ft (approximately 3,600 ft). Then find the density altitude at 60°F and 1,750 ft pressure altitude (approximately 2,300 ft). Thus, the density altitude decreases 1,300 ft (3,600 ft - 2,300 ft).

Chapter 3: Airplane Performance

Refer to Figure 1 on page 36 for questions 13 and 14.

13.
0027. (Refer to figure 1.) Determine the density altitude for these conditions.

Altimeter setting 29.25
Rwy temperature .... +81 °F
Airport elevation .... 5,250 ft

A— 4,600 feet.
B— 5,877 feet.
C— 7,700 feet.
D— 8,400 feet.

Answer (D) is correct (0027). *(PHAK Chap IV)*
With an altimeter setting of 29.25" Hg, about 626 ft (579 plus ½ the 94 ft pressure altitude conversion factor difference between 29.2 and 29.3) must be added to the field elevation of 5,250 ft to obtain the pressure altitude, or 5,876 ft. On the chart, find the point at which the pressure altitude line for 5,876 ft crosses the 81°F line. The density altitude at that spot shows somewhere in the mid-8000s ft. The closest answer choice is 8,400 ft.

14.
0028. (Refer to figure 1.) Determine the density altitude for these conditions.

Altimeter setting 30.35
Rwy temperature .... +25 °F
Airport elevation .... 3,894 ft

A— 2,000 feet.
B— 2,900 feet.
C— 3,500 feet.
D— 3,800 feet.

Answer (A) is correct (0028). *(PHAK Chap IV)*
With an altimeter setting of 30.35" Hg, 394 ft must be subtracted from a field elevation of 3,894 to obtain a pressure altitude of 3,500 ft. The 394 ft was found by interpolation: 30.3 on the chart is -348, and 30.4 is -440 ft. Adding one-half the -92 ft difference gives -394 ft. Once you have found the pressure altitude, use the chart to plot 3,500 ft pressure altitude at 25°F, to reach 2,000 density altitude. Note that since the temperature is lower than standard, the density altitude is lower than the pressure altitude.

TAKEOFF DISTANCE

15.
0046. (Refer to figure 5.) Determine the total distance required for a takeoff to clear a 50-foot obstacle.

OAT .... Std
Pressure altitude .... 3,000 ft
Headwind component .... Calm

A— 2,000 feet.
B— 2,030 feet.
C— 2,500 feet.
D— 2,800 feet.

Answer (C) is correct (0046). *(PHAK Chap IV)*
Begin on the lower left of Figure 5. Proceed upward on the standard temperature line to the intersection with the 3,000-ft pressure altitude line. From that point, move horizontally to the right to the "over 50-ft barrier" lines. Note there is no wind, so go to the middle (unbroken) line. From that intersection, go vertically down to the "take-off distance" and find 2,500, i.e., halfway between the 2,000 and 3,000 ft lines.

16.
0047. (Refer to figure 5.) What is the approximate ground roll distance required for takeoff?

OAT .... 70 °F
Pressure altitude .... 4,000 ft
Headwind component .... 15 kts

A— 1,500 feet.
B— 2,100 feet.
C— 2,600 feet.
D— 3,200 feet.

Answer (B) is correct (0047). *(PHAK Chap IV)*
Begin on the lower left of Figure 5. Proceed up vertically on the 70°F line to the intersection with the 4,000 ft pressure altitude line. From that intersection, proceed horizontally to the right to the first set of vertical lines which involve "ground roll." Since there is a 15-kt headwind, use the first (leftmost) dashed line. From that intersection, go vertically down to the "take-off distance" scale and you are just to the right of 2,000 ft. The best answer is 2,100 ft.

Chapter 3: Airplane Performance

17.
0048. (Refer to figure 5.) What is the total distance required for a takeoff to clear a 50-foot obstacle?

OAT . . . . Std
Pressure altitude . . . . 5,000 ft
Headwind component . . . . Calm

A— 1,800 feet.
B— 2,100 feet.
C— 2,800 feet.
D— 3,500 feet.

Answer (5) is correct (0048). *(PHAK Chap IV)*
Begin on the lower left of Figure 5. Proceed up the standard temperature line to the intersection of the 5,000 ft pressure altitude line. From that intersection, proceed horizontally to the right over to the second set of vertical lines on the right-hand side of the diagram ("over 50-ft barrier"). Since there is no headwind, use the unbroken middle line and proceed vertically down to the scale at the bottom of the chart, which indicates about 3,100 ft. The FAA will accept all answers as correct to this question.

18.
0049. (Refer to figure 5.) Determine the approximate ground roll distance required for takeoff.

OAT . . . . 90 °F
Pressure altitude . . . . Sea level
Tailwind component . . . . 5 kts

A— 1,400 feet.
B— 1,700 feet.
C— 2,300 feet.
D— 2,700 feet.

Answer (C) is correct (0049). *(PHAK Chap IV)*
Begin on the lower left side of Figure 5. Find the 90°F vertical line, and go up that line to the intersection with "Sea Level." From that intersection, proceed horizontally to the right to the first set of vertical lines which concern ground roll. Note the tailwind component of 5 kts, which requires you to use the second broken line, or the line on the far right. From that intersection, proceed vertically downward to the scale at the bottom of the graph and find approximately 2,300 ft.

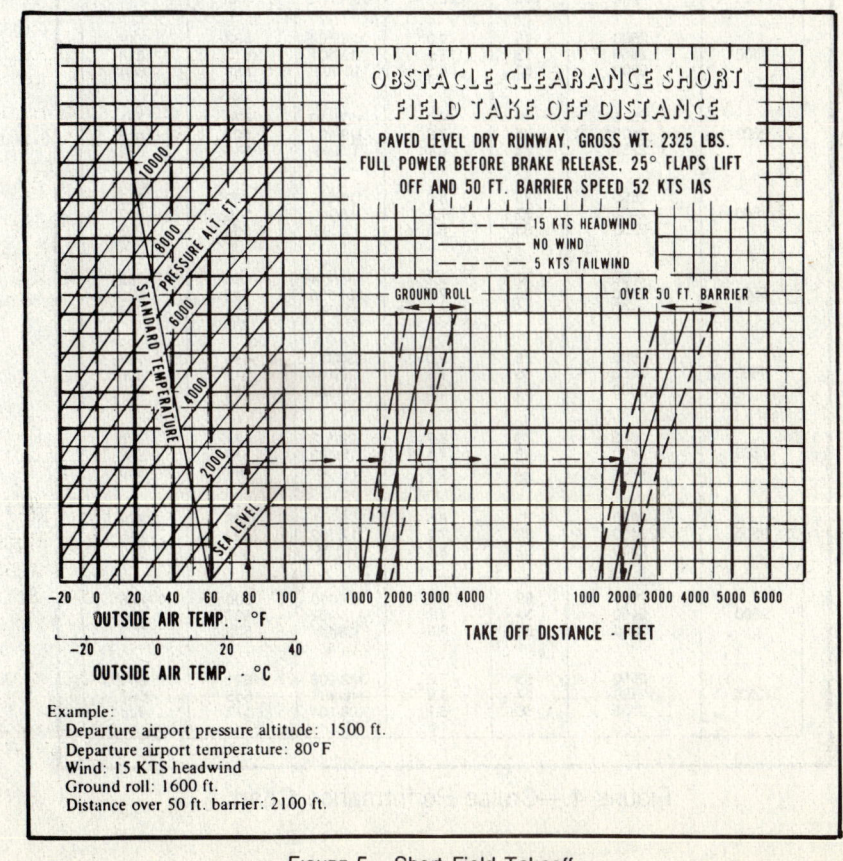

FIGURE 5.—Short Field Takeoff.

CRUISE PERFORMANCE

19.
0041. (Refer to figure 4.) What is the expected fuel consumption for a local flight lasting 4 hours?

Density altitude . . . . 6,500 ft
Throttle setting . . . . 2,450 RPM
Wind . . . . Calm

A— 7.6 gallons.
B— 26.4 gallons.
C— 30.4 gallons.
D— 36.0 gallons.

Answer (C) is correct (0041). *(PHAK Chap IV)*
The requirement is the fuel consumption for 4 hrs. Use Figure 4 to determine the fuel usage in terms of gallons per hour. Begin by going to the first column, "Altitude Feet" and finding the block of information for density altitude of 6,500. In the second column find 2450 RPM. Then proceed horizontally to the right to "Fuel Flow Gal/Hr" in the fourth column, which indicates 7.6 gal/hr. Since the flight is going to last 4 hrs, multiply 4 hrs times 7.6 gal/hr to equal 30.4 gallons of fuel consumed.

CRUISE PERFORMANCE
STANDARD DAY

NOTE: Range includes start, taxi, climb, descent and a 45 minute reserve at 55% maximum continuous power.

ALTITUDE FEET	THROTTLE SETTINGS RPM	% BHP	FUEL FLOW GAL/HR	TAS MPH/KTS	RANGE ST. MILES INITIAL FUEL ONBOARD 38.8	58.8
2500	2500 / 2350 / 2200	75 / 63 / 53	9.0 / 7.6 / 6.5	126/109 / 116/101 / 107/93	469 / 513 / 559	749 / 818 / 894
3500	2525 / 2400 / 2250	75 / 65 / 55	9.0 / 7.8 / 6.7	127/110 / 119/103 / 110/96	470 / 507 / 546	751 / 813 / 874
4500	2550 / 2400 / 2250	75 / 63 / 53	9.0 / 7.6 / 6.5	128/111 / 118/102 / 109/96	472 / 515 / 564	756 / 827 / 905
5500	2600 / 2450 / 2300	77 / 65 / 55	9.2 / 7.8 / 6.7	131/114 / 121/105 / 112/97	468 / 512 / 551	752 / 823 / 887
6500	2600 / 2450 / 2300	75 / 63 / 54	9.0 / 7.6 / 6.6	130/113 / 120/104 / 111/96	475 / 519 / 556	763 / 836 / 897
7500	2600 / 2450 / 2300	73 / 62 / 53	8.7 / 7.4 / 6.5	129/112 / 119/103 / 111/96	481 / 519 / 564	775 / 838 / 910
8500	2600 / 2450 / 2300	71 / 60 / 52	8.5 / 7.2 / 6.4	128/111 / 117/102 / 110/96	488 / 522 / 564	879 / 845 / 913
9500	2600 / 2450 / 2300	69 / 59 / 51	8.3 / 7.1 / 6.3	127/110 / 117/102 / 109/95	500 / 537 / 572	806 / 866 / 923
10500	2550 / 2450 / 2300	63 / 57 / 50	7.6 / 6.9 / 6.1	122/106 / 116/101 / 108/104	521 / 545 / 573	843 / 882 / 928

FIGURE 4.—Cruise Performance Chart.

Chapter 3: Airplane Performance

20.
0042. (Refer to figure 4.) What is the expected fuel consumption for a 3 hour 20 minute flight?

Density altitude .... 3,500 ft
Throttle setting .... 2,250 RPM
Wind .... Calm

A— 18.4 gallons.
B— 22.4 gallons.
C— 25.4 gallons.
D— 30.0 gallons.

Answer (B) is correct (0042). *(PHAK Chap IV)*
The requirement is the fuel consumption for a 3 hr 20 min flight. Use Figure 4 to determine the fuel burn per hour. Begin by locating 3,500 ft density altitude in the first column. In the second column, locate the 2250 RPM line of information. Proceed to the right to the "Fuel Flow Gal/Hr" column, and find 6.7 gal/hr. Multiply 3-1/3 hrs times 6.7 gal/hr to determine approximately 22.4 gallons of fuel consumption.

21.
0043. (Refer to figure 4.) What fuel flow should a pilot expect at 9,500 feet on a standard day with 69 percent BHP?

A— 6.3 gallons per hour.
B— 7.1 gallons per hour.
C— 8.3 gallons per hour.
D— 8.7 gallons per hour.

Answer (C) is correct (0043). *(PHAK Chap IV)*
The requirement is the fuel flow in gal/hr. Begin by finding the 9,500 ft altitude block of information in the first column. Then go to the third column to determine the top line is 69% BHP. Then move to the fourth column and find fuel flow of 8.3 gal/hr.

22.
0044. (Refer to figure 4.) Determine the approximate RPM setting with a standard altitude of 2,500 feet and 63 percent BHP.

A— 2,250 RPM.
B— 2,350 RPM.
C— 2,400 RPM.
D— 2,500 RPM.

Answer (B) is correct (0044). *(PHAK Chap IV)*
The requirement is the RPM setting to obtain 63% BHP at 2,500 ft standard altitude. Find the 2,500 ft block of information in the first column. Go to the third column and find the 63% BHP line. Then move directly to the left to the second column to find the throttle setting of 2,350 RPM.

23.
0045. (Refer to figure 4.) Approximately what true airspeed should a pilot expect with a density altitude of 10,000 feet and 50 percent BHP?

A— 95 knots.
B— 100 knots.
C— 104 knots.
D— 110 knots.

Answer (B) is correct (0045). *(PHAK Chap IV)*
The requirement is the true airspeed at 10,000 ft. Note that blocks of information are provided for 9,500 ft and 10,500 ft (see Column 1, "Altitude Feet"). For both of these blocks of information, move to the right to Column 3, "% BHP" and note that the bottom line of each block of data indicates a percentage of brake horsepower of 51 and 50 respectively. For each of these lines, proceed right to the fifth column, which gives true airspeed in miles and knots, which are 95 kts and 104 kts respectively. Use interpolation to determine that at 10,000 ft density altitude, the true airspeed would be about 100 kts, considering that it would be 95 kts at 9,500 ft, and 104 kts at 10,500 ft.

CROSSWIND COMPONENTS

24.
0029. (Refer to figure 2.) What is the crosswind component for a landing on Rwy 18 if the tower reports the wind as 220° at 30 knots?

A— 19 knots.
B— 23 knots.
C— 30 knots.
D— 34 knots.

25.
0030. (Refer to figure 2.) What is the headwind component for a landing on Rwy 18 if the tower reports the wind as 220° at 30 knots?

A— 19 knots.
B— 23 knots.
C— 30 knots.
D— 34 knots.

Answer (A) is correct (0029). *(PHAK Chap IV)*

The requirement is the crosswind component, which is found on the horizontal axis of the graph. You are given a 30-kt windspeed (the windspeed is shown on the circular lines or arcs). First, calculate the angle between the wind and the runway (220° - 180° = 40°). Next, find the intersection of the 40° line and the 30-kt headwind arc. Then, proceed downward to determine a crosswind component of 19 kts.

Answer (B) is correct (0030). *(PHAK Chap IV)*

The headwind component is on the vertical axis (left-hand side of the graph). Find the same intersection as in the preceding question, i.e., the 30 kt windspeed arc, and the 40° angle between wind direction and flight path (220° - 180°). Then move horizontally to the left and read approximately 23 kts.

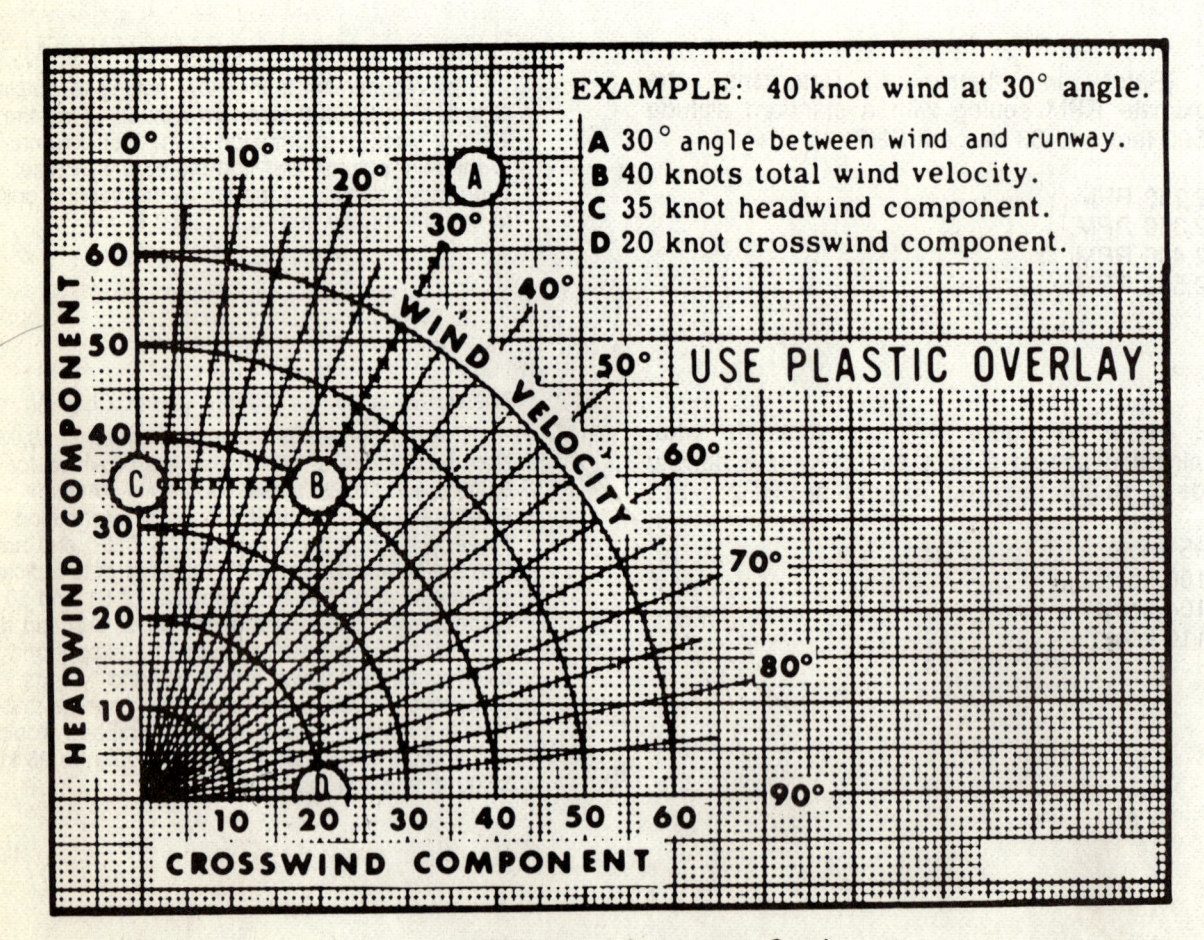

FIGURE 2.—Crosswind Component Graph.

Chapter 3: Airplane Performance 43

26.
0031. (Refer to figure 2.) Determine the maximum wind velocity for a 45° crosswind if the maximum crosswind component for the airplane is 25 knots.

A— 18 knots.
B— 25 knots.
C— 29 knots.
D— 35 knots.

Answer (D) is correct (0031). *(PHAK Chap IV)*
    Start on the bottom of the graph's horizontal axis at 25 kts and move upward to the 45° angle between wind direction and flight path line (halfway between the 40° and 50° lines). Note that you are halfway between the 30 and 40 arc-shaped wind-speed lines, which means that the maximum wind velocity for a 45° crosswind is 35 kts if the airplane is limited to a 25-kt crosswind component.

27.
0032. (Refer to figure 2.) With a reported wind of north at 20 knots, which runway (6, 14, 24, or 32) is appropriate for an airplane with a 13-knot maximum crosswind component?

A— Rwy 6.
B— Rwy 14.
C— Rwy 24.
D— Rwy 32.

Answer (D) is correct (0032). *(PHAK Chap IV)*
    Remember that the runway number indicates the magnetic degrees of the runway by adding a zero, e.g., Runway 6 means a 60° heading. If the wind is from the north (i.e., either 360° or 0°) at 20 kts, Runway 32, i.e., 320°, would provide a 40° crosswind component (360° - 320°) versus Runway 6 providing a 60° crosswind component. Given a 20-kt wind, find the intersection between the 20-kt arc and the angle between wind direction and the flight path of 40°. Dropping straight downward to the horizontal axis gives 13 kts, which is the maximum crosswind component of the example airplane. Runway 6 would have a crosswind component of approximately 17 kts. Note that it is not necessary to compute the crosswind components for the other two runways because you would have a tailwind if you landed on them.

28.
0033. (Refer to figure 2.) What is the maximum wind velocity for a 30° crosswind if the maximum crosswind component for the airplane is 12 knots?

A— 13 knots.
B— 17 knots.
C— 21 knots.
D— 24 knots.

Answer (D) is correct (0033). *(PHAK Chap IV)*
    Start on the graph's horizontal axis at 12 kts and move upward to the 30° angle between wind direction and flight path line. Note that you are almost halfway between the 20 and 30 arc-shaped windspeed lines, which means that the maximum wind velocity for a 30° crosswind is approximately 24 kts if the airplane is limited to a 12-kt crosswind component.

29.
0034. (Refer to figure 2.) With a reported wind of south at 20 knots, which runway (6, 14, 24, or 32) is appropriate for an airplane with a 13-knot maximum crosswind component?

A— Rwy 6.
B— Rwy 14.
C— Rwy 24.
D— Rwy 32.

Answer (B) is correct (0034). *(PHAK Chap IV)*
    If the wind is from the south (i.e., 180°) at 20 kts, Runway 14 (i.e., 140°) would provide a 40° crosswind component (180° - 140°) versus Runway 24, which would provide a 60° crosswind component. On Figure 2, find the intersection between the 20° arc and the angle between the wind direction and the flight path of 40°. Dropping straight down to the horizontal axis gives us 13 kts, which is the maximum crosswind component allowed for this airplane. Runway 24 would have a crosswind component of approximately 17 kts. Note it is not necessary to compute the crosswind components for the other two runways because you would have a tailwind if you landed on them.

Chapter 3: Airplane Performance

LANDING DISTANCE

30.

0050. (Refer to figure 6.) Determine the total distance required to land over a 50-foot obstacle.

OAT .... Std
Pressure altitude .... 4,000 ft
Wind component .... Calm

A— 650 feet.
B— 960 feet.
C— 1,100 feet.
D— 1,190 feet.

Answer (D) is correct (0050). *(PHAK Chap IV)*

Use Figure 6 and begin on the horizontal scale at the lower left. Find the standard temperature line and proceed up to the intersection with the 4,000 ft pressure altitude line. From that intersection, move horizontally to the right to the three lines on the far right, which are for over a 50-ft barrier. Since the wind component is calm, utilize the middle line, which is not broken. From that intersection, move vertically down to the scale at the bottom of the graph. Since you are just to the left of 1200 ft, the landing distance over a 50-ft obstacle is 1,190 ft.

31.

0051. (Refer to figure 6.) Determine the ground roll landing distance required.

OAT .... 90 °F
Pressure altitude .... 3,000 ft
Headwind component .... 15 kts

A— 540 feet.
B— 680 feet.
C— 1,000 feet.
D— 1,200 feet.

Answer (A) is correct (0051). *(PHAK Chap IV)*

Use Figure 6 and begin on the horizontal scale at the lower left. Find the 90°F line and proceed vertically up to the intersection with the 3,000 ft pressure altitude line. From that point, move horizontally to the right to the first set of three vertical lines, which pertain to ground roll. Since the headwind component is 15 kts, use the first line, which is broken. From that intersection, move vertically down to the scale at the bottom of the graph to 540 ft.

32.

0052. (Refer to figure 6.) What total landing distance is required?

OAT .... 20 °F
Pressure altitude .... 7,000 ft
Wind component .... Calm
Obstacle .... 50 ft

A— 700 feet.
B— 1,020 feet.
C— 1,240 feet.
D— 1,470 feet.

Answer (C) is correct (0052). *(PHAK Chap IV)*

Use Figure 6 and begin on the horizontal scale at the lower left. Find the "outside air temperature" line for 20°F and proceed vertically up to the intersection with the 7,000 ft pressure altitude line. From that point, move horizontally to the far right to the set of three vertical lines which pertain to "over 50-ft barrier." Since the wind component is calm, use the middle line, which is unbroken. From that intersection, move vertically down to the scale at the bottom of the graph to 1,240 ft.

33.

0053. (Refer to figure 6.) What approximate ground roll landing distance is required?

OAT .... 30 °F
Pressure altitude .... 6,000 ft
Tailwind component .... 5 kts

A— 680 feet.
B— 880 feet.
C— 1,220 feet.
D— 1,460 feet.

Answer (B) is correct (0053). *(PHAK Chap IV)*

Use Figure 6 and begin on the horizontal scale at the lower left. Find the "outside air temperature" line for 30°F and proceed vertically up to the intersection with the 6,000 ft pressure altitude line. From that point, move horizontally to the right to the first set of three vertical lines which pertain to "ground roll." Since the tailwind component is 5 kts, use the broken line on the right. From that intersection, move vertically down to the scale at the bottom of the graph to 880 ft.

Chapter 3: Airplane Performance

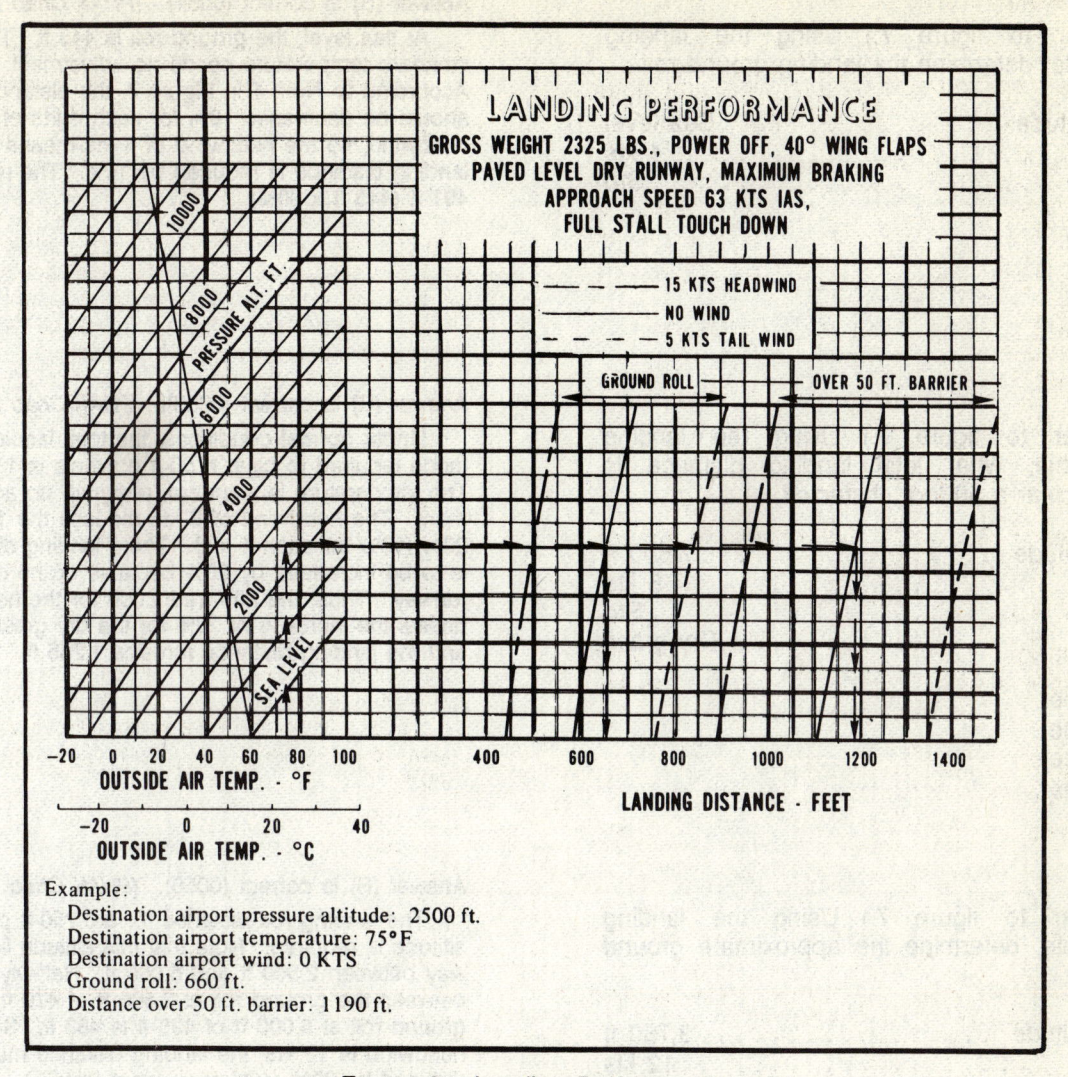

LANDING PERFORMANCE
GROSS WEIGHT 2325 LBS., POWER OFF, 40° WING FLAPS
PAVED LEVEL DRY RUNWAY, MAXIMUM BRAKING
APPROACH SPEED 63 KTS IAS,
FULL STALL TOUCH DOWN

——— 15 KTS HEADWIND
——— NO WIND
- - - - 5 KTS TAIL WIND

Example:
Destination airport pressure altitude: 2500 ft.
Destination airport temperature: 75°F
Destination airport wind: 0 KTS
Ground roll: 660 ft.
Distance over 50 ft. barrier: 1190 ft.

FIGURE 6.—Landing Performance.

46   Chapter 3: Airplane Performance

34.
0054. (Refer to figure 7.) Using the landing distance table, determine the landing ground roll.

Pressure altitude . . . . Sea level
Headwind . . . . 4 kts
Temperature . . . . Std

A— 356 feet.
B— 401 feet.
C— 490 feet.
D— 534 feet.

Answer (B) is correct (0054). *(PHAK Chap IV)*
At sea level, the ground roll is 445 ft. The standard temperature needs no adjustment. According to Note 1 in Figure 7, the distance should be decreased 10% for each 4 kts of headwind, so the headwind of 4 kts means that the landing distance is reduced by 10%. The result is 401 ft (445 ft x 90%).

35.
0055. (Refer to figure 7.) Using the landing distance table, what total landing distance is required to clear a 50-foot obstacle?

Pressure altitude . . . . 7,500 ft
Headwind . . . . 8 kts
Temperature . . . . Std
Runway . . . . Dry grass

A— 1,004 feet.
B— 1,255 feet.
C— 1,506 feet.
D— 1,757 feet.

Answer (B) is correct (0055). *(PHAK Chap IV)*
Under normal conditions, the total landing distance required to clear a 50-ft obstacle is 1,255 ft. The temperature is standard, requiring no adjustment. The headwind of 8 kts reduces the 1,255 by 20% (10% for each 4 kts). Then, landing distance is to be increased by 20% because of the dry grass runway. Thus, the 20% reduction for the headwind offsets the increase of 20% for the dry grass runway and the landing distance remains 1,255 ft.

36.
0056. (Refer to figure 7.) Using the landing distance table, determine the approximate ground roll distance.

Pressure altitude . . . . 3,750 ft
Headwind . . . . 12 kts
Temperature . . . . Std

A— 193 feet.
B— 338 feet.
C— 628 feet.
D— 772 feet.

Answer (B) is correct (0056). *(PHAK Chap IV)*
The landing roll distance for a 3,750-ft pressure altitude is required. Note that this altitude lies halfway between 2,500 ft and 5,000 ft. Halfway between the ground roll at 2,500 ft of 470 ft and the ground roll at 5,000 ft of 495 ft is 483 ft. Since the headwind is 12 kts, the landing distance must be reduced by 30% (10% for each 4 kts):

70% x 483 ft = 338 ft.

37.
0057. (Refer to figure 7.) Using the landing distance table, what total landing distance is required to clear a 50-foot obstacle?

Pressure altitude . . . . 5,000 ft
Headwind . . . . 8 kts
Temperature . . . . 41 °F
Runway . . . . Hard surface

A— 837 feet.
B— 956 feet.
C— 1,076 feet.
D— 1,554 feet.

Answer (B) is correct (0057). *(PHAK Chap IV)*
Under standard conditions, the distance to land over a 50-ft obstacle at 5,000 ft is 1,195 ft. The temperature is standard, requiring no adjustment. The headwind of 8 kts, however, requires that the distance be decreased by 20% (10% for each 4 kts headwind). Thus, the landing ground roll will be 956 ft (80% of 1,195).

Chapter 3: Airplane Performance

LANDING DISTANCE

FLAPS LOWERED TO 40° - POWER OFF
HARD SURFACE RUNWAY - ZERO WIND

GROSS WEIGHT LBS.	APPROACH SPEED, IAS, MPH	AT SEA LEVEL & 59° F.		AT 2500 FT. & 50° F.		AT 5000 FT. & 41° F.		AT 7500 FT. & 32° F.	
		GROUND ROLL	TOTAL TO CLEAR 50 FT. OBS	GROUND ROLL	TOTAL TO CLEAR 50 FT. OBS	GROUND ROLL	TOTAL TO CLEAR 50 FT. OBS	GROUND ROLL	TOTAL TO CLEAR 50 FT. OBS
1600	60	445	1075	470	1135	495	1195	520	1255

NOTES:
1. Decrease the distances shown by 10% for each 4 knots of headwind.
2. Increase the distance by 10% for each 60° F. temperature increase above standard.
3. For operation on a dry, grass runway, increase distances (both "ground roll" and "total to clear 50 ft. obstacle") by 20% of the "total to clear 50 ft. obstacle" figure.

FIGURE 7.—Airplane Landing Distance Table.

38.
0058. (Refer to figure 7.) Using the landing distance table, determine the total distance required to land over a 50-foot obstacle.

Pressure altitude . . . . 5,000 ft
Headwind . . . . Calm
Temperature . . . . 101 °F

A— 239 feet.
B— 1,099 feet.
C— 1,291 feet.
D— 1,314 feet.

Answer (D) is correct (0058). *(PHAK Chap IV)*
The normal distance required to clear a 50-ft obstacle at 5,000 ft is 1,195 ft. Since the temperature is 60°F above standard, the distance should be increased by 10%:

1,195 ft x 110% = 1,314 ft.

39.
0059. (Refer to figure 7.) Using the landing distance table, what is the approximate landing ground roll distance?

Pressure altitude . . . . 1,250 ft
Headwind . . . . 8 kts
Temperature . . . . Std

A— 275 feet.
B— 366 feet.
C— 470 feet.
D— 549 feet.

Answer (B) is correct (0059). *(PHAK Chap IV)*
The landing ground roll at a pressure altitude of 1,250 ft is required. The difference between landing distance at sea level and 2,500 ft is 25 ft (470 - 445). One-half of this distance (12 ft) plus the 445 ft at sea level is 457 ft. The temperature is standard, requiring no adjustment. The headwind of 8 kts requires the distance to be decreased by 20%. Thus, the distance required will be 366 ft (457 x 80%).

WINGTIP VORTICES

40.
0242. Wingtip vortices created by large aircraft tend to

A— sink below the aircraft generating the turbulence.
B— rise into the traffic pattern.
C— rise into the takeoff or landing path of a crossing runway.
D— accumulate at the beginning of the takeoff roll.

Answer (A) is correct (0242). *(AIM Para 543)*
Wingtip vortices created by large airplanes tend to sink below the airplane generating the turbulence.
Answers (B) and (C) are incorrect because wingtip vortices do not rise or gain altitude, but sink toward the ground. However, they may move horizontally left or right depending on crosswind conditions. Answer (D) is incorrect because wingtip vortices begin to occur as the airplane develops lift just prior to or at the point of takeoff. A taxiing airplane does not produce wingtip vortices, although it does produce prop blast or jet thrust turbulence.

41.
0241. Wingtip vortices, the dangerous turbulence that might be encountered behind a large aircraft, are created only when that aircraft is

A— operating at high airspeeds.
B— heavily loaded.
C— developing lift.
D— using high-power settings.

Answer (C) is correct (0241). *(AIM Paras 541, 543)*
Wingtip vortices, a wake turbulence that can be encountered behind a large airplane, are created only when that airplane is developing lift. Wingtip vortices do not develop when a large airplane is taxiing, although prop blast or jet thrust turbulence can be experienced near the rear of a taxiing large airplane.

Answer (A) is incorrect because the greatest turbulence is produced from an airplane operating at a slower airspeed. Answer (B) is incorrect because, even though a heavily loaded airplane may produce greater turbulence, an airplane does not have to be heavily loaded in order to produce wingtip vortices. Wingtip vortices are produced whenever the airplane is developing lift. Answer (D) is incorrect because an airplane does not have to be using a high power setting to produce wingtip vortices or severe turbulence. The greatest turbulence exists when the airplane is producing maximum lift.

42.
0245. When taking off or landing at a busy airport where large, heavy aircraft are operating, one should be particularly alert to the hazards of wingtip vortices because this turbulence tends to

A— rise from a crossing runway into the takeoff or landing path.
B— rise into the traffic pattern area surrounding the airport.
C— sink into the flightpath of aircraft operating below the aircraft generating the turbulence.
D— accumulate at the beginning of the takeoff roll.

Answer (C) is correct (0245). *(AIM Para 545)*
When taking off or landing at a busy airport where large, heavy airplanes are operating, you should be particularly alert to the hazards of wingtip vortices because this turbulence tends to sink into the flightpaths of airplanes operating below the airplane generating the turbulence. Wingtip vortices are caused by a differential in high and low pressure at the wingtip of an airplane, creating a spiraling effect trailing behind the wingtip, similar to a horizontal tornado.

Answers (A) and (B) are incorrect because wingtip vortices always trail behind an airplane and descend toward the ground. However, they do drift with the wind and will not stay directly behind an airplane if there is a crosswind. Answer (D) is incorrect because these turbulences begin just prior to liftoff, not at the beginning of the takeoff roll.

43.
0498. What wind condition prolongs the hazards of wake turbulence on a landing runway for the longest period of time?

A— Direct headwind.
B— Direct tailwind.
C— Light quartering tailwind.
D— Light quartering headwind.

Answer (C) is correct (0498). *(AIM Para 543)*
The most dangerous wind condition when avoiding wake turbulence on landing is a light, quartering tailwind. The tailwind can push the vortices forward which could put it in the touchdown zone of your aircraft even if you used proper procedures and landed beyond the touchdown point of the preceding aircraft. Also the quartering wind may push the upwind vortices to the middle of the runway.

Answers (A) and (D) are incorrect because headwinds push the vortices out of your touchdown zone if you land beyond the touchdown point of the preceding aircraft. Answer (B) is incorrect because a direct tailwind would permit the vortices to move to each side of the runway.

CHAPTER FOUR
AIRPLANE INSTRUMENTS, ENGINES, AND SYSTEMS

```
Compass Turning Errors ........................... (6 questions) ...... 49, 54
Airspeed Indicator ................................ (11 questions) ...... 50, 55
Altimeter ........................................ (6 questions) ...... 50, 59
Types of Altitude ................................ (9 questions) ...... 51, 60
Altimeter Calculations ........................... (2 questions) ...... 51, 62
Altimeter Errors ................................. (5 questions) ...... 51, 62
Engine Temperature .............................. (6 questions) ...... 52, 64
Constant-Speed Propeller ........................ (3 questions) ...... 52, 65
Engine Ignition Systems ......................... (2 questions) ...... 52, 66
Carburetor Icing ................................. (6 questions) ...... 52, 67
Carburetor Heat ................................. (3 questions) ...... 53, 69
Fuel/Air Mixture ................................. (3 questions) ...... 53, 69
Abnormal Combustion ............................ (4 questions) ...... 53, 70
Aviation Fuel Practices .......................... (4 questions) ...... 53, 72
```

This chapter contains outlines of major concepts tested, all FAA test questions and answers regarding the major mechanical and instrument systems in an airplane, and an explanation of each answer. The subtopics or modules within this chapter are listed above, followed in parentheses by the number of questions from the FAA written test pertaining to that particular module. The two numbers following the parentheses are the page numbers on which the outline and questions begin for that module.

CAUTION: Recall that the <u>sole purpose</u> of this book is to expedite your passing the FAA written test for the recreational pilot certificate. Accordingly, all extraneous material (i.e., topics or regulations not directly tested on the FAA written test) is omitted, even though much more information and knowledge are necessary to fly safely. This additional material is presented in *RECREATIONAL PILOT FLIGHT MANEUVERS* and *PRIVATE PILOT HANDBOOK*, available from Aviation Publications, Inc. See pages 208 to 210 for more information and an order form.

COMPASS TURNING ERRORS (6 questions)

1. During flight, magnetic compasses can only be considered accurate during straight-and-level flight at constant airspeed.

2. The difference between direction indicated by a magnetic compass not installed in an airplane and one installed in an airplane is called **deviation**.

    a. Magnetic fields produced by metals and electrical accessories in an airplane disturb the compass needles.

3. In the northern hemisphere, **acceleration/deceleration error** occurs when on an **easterly or westerly heading**.

    a. A magnetic compass will indicate a **turn toward the north during acceleration** when on an easterly or westerly heading.

    b. A magnetic compass will indicate a **turn toward the south during deceleration** when on an easterly or westerly heading.

c. Acceleration/deceleration error does not occur when on a northerly or southerly heading.

4. In the northern hemisphere, **compass turning error** occurs **when turning from a northerly or southerly heading**.
    a. A magnetic compass **will lag** (and at the start of a turn indicate a turn in the opposite direction) **when turning from a northerly heading**.
        1) If turning to the east (right), the compass will initially indicate west and then lag behind the actual heading until your airplane is headed east (at which point there is no error).
        2) If turning to the west (left), the compass will initially indicate east and then lag behind the actual heading until your airplane is headed west (at which point there is no error).
    b. The compass **will lead** or precede the turn **when turning from a southerly heading**.
    c. Turning errors do not occur when turning from an easterly or westerly heading.

5. These errors diminish as acceleration, deceleration, or turns are completed.

AIRSPEED INDICATOR (11 questions)

1. Airspeed indicators have several color-coded markings.
    a. The white arc is the **flap operating range**.
        1) The lower limit is the **power-off stalling speed** with wing flaps and landing gear in the landing position.
        2) The upper limit is the **maximum flaps-extended speed**.
    b. The green arc is the **normal operating range**.
        1) The lower limit is the power-off **stalling speed with the wing flaps up and landing gear retracted** (called a specified configuration).
        2) The upper limit is the **maximum structural cruising speed for normal operation**.
    c. The yellow arc is airspeed which is safe in smooth air only.
        1) It is known as the caution range.
    d. The red line is the speed that should never be exceeded (i.e., the "never exceed speed").

2. The most important airspeed limitation that is not color-coded is the maneuvering speed.
    a. The *maneuvering speed* is the maximum speed at which full deflection of aircraft controls can be made without causing structural damage.
    b. It is the maximum speed for flight in **turbulent air**.

ALTIMETER (6 questions)

1. Altimeters have three "hands" (e.g., as a clock has the hour, minute, and second hands).
2. The **three hands** on the altimeter are the
    a. 10,000-ft interval (short needle).
    b. 1,000-ft interval (medium needle).
    c. 100-ft interval (long needle).

Chapter 4: Airplane Instruments, Engines, and Systems

3. Altimeters are numbered 0-9.
4. To read an altimeter,
    a. First, determine whether the short needle points between 0 and 1 (1-10,000), 1-2 (10,000-20,000), or 2-3 (20,000-30,000).
    b. Second, determine whether the medium needle is between 0 and 1 (0-1,000), 1 and 2 (1,000-2,000), etc.
    c. Third, determine at which number the long needle is pointing, e.g., 1 for 100 ft, 2 for 200 ft, etc.
5. Prior to takeoff, the altimeter should be set to the **current local altimeter setting**.
    a. If the current local altimeter setting is not available, use the **departure airport elevation**.

TYPES OF ALTITUDE (9 questions)

1. *Absolute altitude* is the altitude above the surface.
2. *True altitude* is the actual distance above mean sea level. It is not susceptible to variation with atmospheric conditions.
3. *Density altitude* is pressure altitude corrected for temperatures other than standard.
4. *Pressure altitude* is the indicated altitude when the altimeter setting is adjusted to 29.92 inches of mercury (also written 29.92" Hg).
    a. Pressure altitude is indicated altitude corrected for nonstandard pressure.
    b. Pressure altitude and density altitude are thus the same at standard temperatures.
5. *Indicated altitude* is the same as true altitude when standard conditions exist and the altimeter is calibrated appropriately.
6. When the altimeter is adjusted on the ground so that indicated altitude equals airport elevation, the altimeter setting is that for your location, i.e., approximately the setting you would get from the control tower.

ALTIMETER CALCULATIONS (2 questions)

1. Atmospheric pressure decreases about 1" Hg for every 1,000 ft of altitude gained.
    a. **Changing the altimeter setting** changes the indicated altitude in the same direction and by 1,000 ft for every inch change.
    b. For example, changing from 29.91 to 30.91 increases indicated altitude by 1,000 ft.

ALTIMETER ERRORS (5 questions)

1. Since altimeter readings are adjusted for changes in barometric pressure but not for temperature changes, an airplane will be at **lower than indicated altitude when flying in colder** than standard temperature air when maintaining a constant indicated altitude.
    a. On warm days, the altimeter **indicates lower than actual altitude**.
2. Likewise, when pressure lowers en route at a constant indicated altitude, your altimeter will indicate higher than actual altitude until you adjust it.

ENGINE TEMPERATURE (6 questions)

1. Excessively high engine temperature either in the air or on the ground will cause loss of power, excessive oil consumption, and excessive wear on the internal engine.
2. An engine is cooled in part by circulating oil through the system to reduce friction and absorb heat from internal engine parts.
3. Engine oil and cylinder head temperatures can exceed their normal operating range because of (among other causes)
   a. Operating with too much power.
   b. Climbing too steeply (at too low an airspeed) in hot weather.
   c. Using fuel that has a lower-than-specified rating.
   d. Operating with too lean a mixture.
   e. The oil level being too low.
4. Excessively high engine temperatures can be reduced by reversing each of the above actions, e.g., reducing power, climbing less steeply (increasing airspeed), using higher octane fuel, enriching the mixture, etc.

CONSTANT-SPEED PROPELLER (3 questions)

1. The advantage of a constant-speed propeller (also known as controllable-pitch) is that it permits the pilot to select the blade angle for the most efficient performance.
2. Constant-speed propeller airplanes have both throttle and propeller controls.
   a. The throttle controls power output, which is registered on the manifold pressure gauge.
   b. The propeller control regulates engine revolutions per mintue (RPM), which are registered on the tachometer.
3. To avoid overstressing cylinders, high manifold pressure should not be used with low RPM settings.

ENGINE IGNITION SYSTEMS (2 questions)

1. One purpose of the dual-ignition system is to provide for improved engine performance.
   a. The other is increased safety.
2. If an engine runs after the ignition is turned off, the magneto groundwire is probably broken.

CARBURETOR ICING (6 questions)

1. Carburetor-equipped engines are more susceptible to icing than fuel-injected engines.
   a. The operating principle of float-type carburetors is a difference in air pressure at the venturi throat and the air inlet.
   b. Fuel-injected engines do not have a carburetor.
2. The first indication of carburetor ice on airplanes with fixed-pitch propellers and float-type carburetors is a loss of RPM.
3. Carburetor ice is likely to form when outside air temperature is between 20°F and 70°F and there is visible moisture or high humidity.

Chapter 4: Airplane Instruments, Engines, and Systems

4. When carburetor heat is applied to eliminate carburetor ice in an airplane equipped with a fixed-pitch propeller, there will be a further decrease in RPM (due to the melting ice) followed by a gradual increase in RPM.

CARBURETOR HEAT (3 questions)

1. Carburetor heat enriches the fuel/air mixture,
    a. Because warm air is less dense than cold air.
    b. When the air density decreases (because the air is warm), the fuel/air mixture (ratio) increases since there is less air for the same amount of fuel.
2. Applying carburetor heat decreases engine output and increases operating temperature.

FUEL/AIR MIXTURE (3 questions)

1. At higher altitudes, the fuel/air mixture must be leaned to decrease the fuel flow in order to compensate for the decreased air density, i.e., to keep the fuel/air mixture constant.
    a. If you descend from high altitudes to lower altitudes without enriching the mixture, the mixture will become lean because the air is more dense at lower altitudes.
2. If you are running up your engine at a high-altitude airport, you may eliminate engine roughness by leaning the mixture.
    a. Particularly if the engine runs even worse with carburetor heat because warm air further enriches the mixture.

ABNORMAL COMBUSTION (4 questions)

1. *Detonation* occurs when the fuel/air mixture explodes, instead of burning evenly.
2. Detonation is usually caused by using a lower-than-specified grade of aviation fuel or by excessive engine temperature.
    a. This causes many engine problems including excessive wear and higher than normal operating temperatures.
3. Retard the throttle if there is suspicion that an engine (with a fixed-pitch propeller) is detonating during climb-out after takeoff.
4. *Preignition* is the uncontrolled firing of the fuel/air charge in advance of the normal spark ignition.

AVIATION FUEL PRACTICES (4 questions)

1. Use of the next-higher-than-specified grade of fuel is better than using the next-lower-than-specified grade of fuel. This will prevent the possibility of detonation, or running the engine too hot.
2. Filling the fuel tanks at the end of the day prevents moisture condensation by eliminating the airspace in the tanks.
3. In an airplane equipped with fuel pumps, running a fuel tank dry before switching tanks while in the air may be disastrous because the fuel pump may draw air into the fuel system and cause vapor lock.

Chapter 4: Airplane Instruments, Engines, and Systems

> **QUESTIONS AND ANSWER EXPLANATIONS**
>
> All of the FAA questions from the written test for the Recreational Pilot certificate relating to airplane instruments, engines, and systems and the material outlined previously are reproduced below in the same modules as the previous outlines. To the immediate right of each question are the correct answer and answer explanation. You should cover these answers and answer explanations while responding to the questions. Refer to the general discussion in Chapter 1 on how to take the examination.
>
> Remember that the questions from the FAA's Written Test Book have been reordered by topic, and the topics have been organized into a meaningful sequence. Accordingly, the first line of the answer explanation gives the FAA question number and the citation of the authoritative source for the answer.

COMPASS TURNING ERRORS

1.
0091. In the Northern Hemisphere, a magnetic compass will normally indicate a turn toward the north if

A— a right turn is entered from an east heading.
B— a left turn is entered from a west heading.
C— an aircraft is accelerated while on an east or west heading.
D— an aircraft is decelerated while on an east or west heading.

Answer (C) is correct (0091). *(PHAK Chap III)*

In the northern hemisphere, a magnetic compass will normally indicate a turn toward the north if an airplane is accelerated while on an east or west heading.

Answers (A) and (B) are incorrect because on turns to the south, the compass will lag behind the turn. Answer (D) is incorrect because, if an airplane is decelerated while on an east or west heading, the compass would show a false turn to the south (i.e., the opposite of the accelerating error).

2.
0093. Deviation in a magnetic compass is caused by

A— presence of flaws in the permanent magnets of the compass.
B— the difference in the location between true north and magnetic north.
C— magnetic ore deposits in the Earth distorting the lines of magnetic force.
D— magnetic fields within the aircraft distorting the lines of magnetic force.

Answer (D) is correct (0093). *(PHAK Chap III)*

The difference between the direction indicated by a magnetic compass not installed in an airplane and one installed in the airplane is called deviation. Magnetic fields produced by metals and electrical accessories in the airplane disturb the compass needle and produce errors.

Answer (A) is incorrect because flaws in the magnets would simply make the compass inaccurate. Answers (B) and (C) are incorrect because they cause compass variation, not deviation.

3.
0092. In the Northern Hemisphere, if an aircraft is accelerated or decelerated, the magnetic compass will normally indicate

A— a turn momentarily, with changes in airspeed on any heading.
B— correctly when on a north or south heading while either accelerating or decelerating.
C— a turn toward the south while accelerating on a west heading.
D— a turn toward the north while decelerating on an east heading.

Answer (B) is correct (0092). *(PHAK Chap III)*

Acceleration and deceleration errors on magnetic compasses do not occur when on a north or south heading in the northern hemisphere. They occur on east and west headings.

Answer (A) is incorrect because acceleration and deceleration errors occur only on easterly and westerly headings. Answers (C) and (D) are incorrect because the reverse is true; i.e., a turn to the north is indicated upon acceleration and a turn to the south is indicated on deceleration on east or west headings.

Chapter 4: Airplane Instruments, Engines, and Systems

4.
0095. In the Northern Hemisphere, a magnetic compass will normally indicate initially a turn toward the west if

A— an aircraft is accelerated while on a north heading.
B— an aircraft is decelerated while on a south heading.
C— a left turn is entered from a north heading.
D— a right turn is entered from a north heading.

Answer (D) is correct (0095). *(PHAK Chap III)*
Due to the northerly turn error in the northern hemisphere, a magnetic compass will initially indicate a turn toward the west if a right turn (east) is entered from a north heading.
Answers (A) and (B) are incorrect because acceleration/deceleration errors do not occur on north and south headings. Answer (C) is incorrect because, if a left turn (west) were made from a north heading, the compass would initially indicate a turn toward the east.

5.
0094. In the Northern Hemisphere, the magnetic compass will normally indicate a turn toward the south when

A— a left turn is entered from an east heading.
B— a right turn is entered from a west heading.
C— the aircraft is accelerated while on an east heading.
D— the aircraft is decelerated while on a west heading.

Answer (D) is correct (0094). *(PHAK Chap III)*
In the northern hemisphere, a magnetic compass will normally indicate a turn toward the south if an airplane is decelerated while on an east or west heading.
Answers (A) and (B) are incorrect because turning errors do not occur from east and west headings. Answer (C) is incorrect because acceleration errors are to the north, not south.

6.
0096. In the Northern Hemisphere, a magnetic compass will normally indicate initially a turn toward the east if

A— an aircraft is accelerated while on a north heading.
B— an aircraft is decelerated while on a south heading.
C— a right turn is entered from a north heading.
D— a left turn is entered from a north heading.

Answer (D) is correct (0096). *(PHAK Chap III)*
In the northern hemisphere, a magnetic compass normally initially indicates a turn toward the east if a left turn (to the west) is entered from a north heading.
Answers (A) and (B) are incorrect because a(n) acceleration/deceleration error is not apparent while on a north or south heading, only on east or west headings. Answer (C) is incorrect because, if a right turn is made (i.e., toward the east) on a north heading, the compass would initially indicate a turn toward the west.

AIRSPEED INDICATOR

7.
0099. The red line on an airspeed indicator means a maximum airspeed that

A— may be exceeded only if gear and flaps are retracted.
B— may be exceeded if abrupt maneuvers are not attempted.
C— may be exceeded only in smooth air.
D— should not be exceeded.

Answer (D) is correct (0099). *(PHAK Chap III)*
The red line on an airspeed indicator indicates the maximum airspeed at which an airplane may be operated (also known as the never-exceed speed).
Answer (A) is incorrect because the upper limit of the white arc is the airspeed above which flaps cannot be extended. Answer (B) is incorrect because the maneuvering speed is the airspeed above which turbulent air should not be penetrated or abrupt maneuvers attempted. Answer (C) is incorrect because the upper limit of the green arc (maximum structural cruising speed) is the airspeed above which an airplane should be operated only in smooth air.

Chapter 4: Airplane Instruments, Engines, and Systems

8.
0111. What is an important airspeed limitation that is not color coded on airspeed indicators?

A— Never-exceed speed.
B— Maximum structural cruising speed.
C— Maneuvering speed.
D— Maximum flaps-extended speed.

Answer (C) is correct (0111). *(PHAK Chap III)*
The maneuvering speed of an airplane is an important airspeed limitation not color-coded on the airspeed indicator. It is found in the airplane manual or placarded on the airplane instrument panel. Maneuvering speed is the maximum speed at which full deflection of the airplane controls can be made without incurring structural damage. Maneuvering speed or less should be held in turbulent air to prevent structural damage due to excessive loads.

Answer (A) is incorrect because the never-exceed speed is indicated on the airspeed indicator by a red line. Answer (B) is incorrect because the maximum structural cruising speed is indicated by the upper limit of the green arc. Answer (D) is incorrect because the maximum flaps-extended speed is indicated by the upper limit of the white arc.

9.
0098. (Refer to figure 9.) Note the color-coded markings on the airspeed indicator. What is the caution range of the airplane?

A— 0 to 60 MPH.
B— 60 to 100 MPH.
C— 100 to 165 MPH.
D— 165 to 208 MPH.

Answer (D) is correct (0098). *(PHAK Chap III)*
The "caution range" is indicated by the yellow color arc on the illustrated airspeed indicator. This is from 165 to 208 mph. Operation within this range is safe only in smooth air.

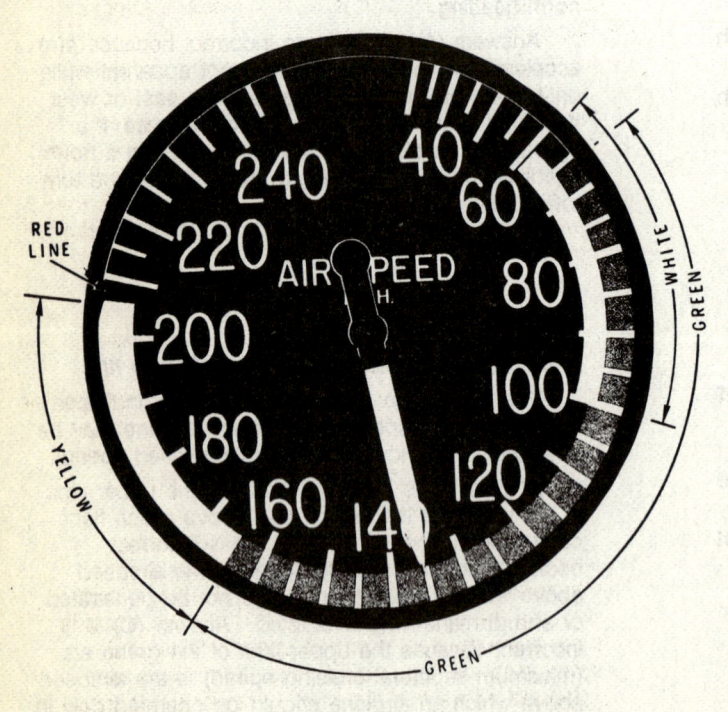

Figure 9.-Airspeed Indicator.

Chapter 4: Airplane Instruments, Engines, and Systems

10.
0100. (Refer to figure 9.) The maximum speed at which the airplane can be operated in smooth air is

A— 65 MPH.
B— 100 MPH.
C— 165 MPH.
D— 208 MPH.

Answer (D) is correct (0100). *(PHAK Chap III)*
The maximum speed at which this airplane can be operated in smooth air is 208 mph, the upper limit of the yellow arc, or the red line. Flight up to this airspeed range should only be in smooth air. Normal operation is usually limited to the green arc (165 mph).

Answer (A) is incorrect because 65 mph is the power-off stall speed with the wing flaps up and the landing gear retracted, at the lower limit of the green arc. Answer (B) is incorrect because 100 mph is the maximum flaps-extended speed, at the upper limit of the white arc. Answer (C) is incorrect because 165 mph is the maximum structural cruising speed, at the upper limit of the green arc.

11.
0097. (Refer to figure 9.) Note the color-coded markings on the airspeed indicator. What is the normal flap operating range for the airplane?

A— 60 to 100 MPH.
B— 60 to 208 MPH.
C— 65 to 165 MPH.
D— 165 to 208 MPH.

Answer (A) is correct (0097). *(PHAK Chap III)*
The normal flap operating range is indicated by the white arc (60 to 100 mph).

Answer (B) is incorrect because 60 to 208 mph is the entire operating range of this airplane. Answer (C) is incorrect because 65 to 165 mph is the normal operating range for this airplane. Answer (D) is incorrect because 165 to 208 mph is the caution range of this airplane.

12.
0101. (Refer to figure 9.) Which of the color-coded markings on the airspeed indicator identifies the never-exceed speed?

A— Upper A/S limit of the white arc.
B— Upper A/S limit of the green arc.
C— Lower A/S limit of the yellow arc.
D— The red radial line.

Answer (D) is correct (0101). *(PHAK Chap III)*
The never-exceed speed is indicated by a red line and is found at the upper limit of the yellow arc. Operating above this speed may result in structural damage.

Answer (A) is incorrect because the upper limit of the white arc is the maximum speed at which flaps may be extended. Answer (B) is incorrect because the upper limit of the green arc is the maximum structural cruising speed, to be exceeded only in smooth air. Answer (C) is incorrect because the lower limit of the yellow arc is the beginning of the caution range.

13.
0102. (Refer to figure 9.) Which color-coded marking identifies the power-off stalling speed with flaps and landing gear in the retracted position?

A— Upper A/S limit of the green arc.
B— Upper A/S limit of the white arc.
C— Lower A/S limit of the green arc.
D— Lower A/S limit of the white arc.

Answer (C) is correct (0102). *(PHAK Chap III)*
The lower airspeed limit of the green arc indicates the power-off stalling speed with the flaps up and landing gear retracted. The lower airspeed limit of the white arc indicates the power-off stalling speed with wing flaps and landing gear in the landing position (down).

Answer (A) is incorrect because the upper airspeed limit of the green arc is the maximum structural cruising speed. Answer (B) is incorrect because the upper airspeed limit of the white arc is the maximum flaps-extended speed. Structural damage to the flaps could occur if the flaps are extended above this airspeed. Answer (D) is incorrect because the lower limit of the white arc is the stalling speed with the flaps and gear extended, not retracted.

Questions 14 through 17 refer to Figure 9, which is presented on page 56.

14.
0104. (Refer to figure 9.) Which of the color-coded markings identifies the normal flap operating range?

A— The lower limit of the white arc to the upper limit of the green arc.
B— The green arc.
C— The white arc.
D— The yellow arc.

Answer (C) is correct (0104). *(PHAK Chap III)*
The normal flap operating range is indicated by the white arc. The power-off stall speed with flaps extended is at the lower limit of the arc, and the maximum speed at which flaps can be extended without damage to them is the upper limit of the arc.

15.
0103. (Refer to figure 9.) What is the maximum flaps-extended speed?

A— 60 MPH.
B— 65 MPH.
C— 100 MPH.
D— 165 MPH.

Answer (C) is correct (0103). *(PHAK Chap III)*
The airspeed indicator, as illustrated, depicts a maximum flaps-extended speed of 100 mph. This is the upper limit of the white arc.
Answer (A) is incorrect because 60 mph is the lower limit of the white arc, which is the power-off stalling speed with flaps and gear in the landing position (down). Answer (B) is incorrect because 65 mph is the lower limit of the green arc, which is the power-off stall speed with the flaps up and landing gear retracted. Answer (D) is incorrect because 165 mph is the upper limit of the green arc, which is the maximum structural cruising speed.

16.
0105. (Refer to figure 9.) Which of the color-coded markings identifies the power-off stalling speed with wing flaps and landing gear in the landing position?

A— Upper A/S limit of the green arc.
B— Upper A/S limit of the white arc.
C— Lower A/S limit of the green arc.
D— Lower A/S limit of the white arc.

Answer (D) is correct (0105). *(PHAK Chap III)*
The lower airspeed limit of the white arc indicates the power-off stalling speed with wing flaps and landing gear in the landing position.
Answer (A) is incorrect because the upper airspeed limit of the green arc is the maximum structural cruising speed. Answer (B) is incorrect because the upper airspeed limit of the white arc is the maximum flaps-extended speed. Structural damage to the flaps could occur if the flaps are extended above this airspeed. Answer (C) is incorrect because the lower airspeed limit of the green arc is the power-off stall speed with the flaps up and landing gear retracted.

17.
0106. (Refer to figure 9.) What is the maximum structural cruising speed?

A— 65 MPH.
B— 100 MPH.
C— 165 MPH.
D— 208 MPH.

Answer (C) is correct (0106). *(PHAK Chap III)*
The airspeed indicator illustration depicts a maximum structural cruising speed of 165 mph as indicated by the upper limit of the green arc. Operation above this airspeed should be done in smooth air only.
Answer (A) is incorrect because 65 MPH is the lower limit of the green arc which indicates stalling speed with gear and flaps retracted. Answer (B) is incorrect because 100 mph is the upper limit of the white arc and is the maximum speed at which the flaps can be extended. Answer (D) is incorrect because 208 mph is the speed that should never be exceeded. Beyond this speed, structural damage to the airplane may occur.

Chapter 4: Airplane Instruments, Engines, and Systems

ALTIMETER

18.
0108. (Refer to figure 10.) Altimeter B indicates

A— 1,500 feet.
B— 4,500 feet.
C— 14,500 feet.
D— 15,500 feet.

Answer (C) is correct (0108). *(PHAK Chap III)*

Altimeter B indicates 14,500 ft because the shortest needle is between the 1 and the 2, indicating about 15,000 ft; the middle needle is between 4 and 5, indicating 4,500 ft; and the long needle is on 5, indicating 500 ft, i.e., 14,500 ft.

Answers (A) and (B) are incorrect because the short needle is between 1 and 2, i.e., over 10,000 ft. Answer (D) is incorrect because the middle needle would have to be pointing between the 5 and 6 to indicate 15,500 ft.

19.
0107. (Refer to figure 10.) Altimeter A indicates

A— 500 feet.
B— 1,500 feet.
C— 10,500 feet.
D— 15,000 feet.

Answer (C) is correct (0107). *(PHAK Chap III)*

The altimeter has three needles. The short needle indicates 10,000-ft intervals, the middle-length needle indicates 1,000-ft intervals, and the long needle indicates 100-ft intervals. In A, the shortest needle is on 1, which indicates about 10,000 ft. The middle-length needle indicates half-way between zero and 1, which is 500 ft. This is confirmed by the longest needle on 5, which indicates the altimeter setting of 10,500 ft.

Answer (A) is incorrect because, if it were indicating just 500 ft, the short and medium needles would have to be on or near zero. Answer (B) is incorrect because, if it were 1,500 ft, the shortest needle would be near zero and the middle needle would be between the 1 and the 2. Answer (D) is incorrect because 15,000 ft would be indicated by the short needle between the 1 and the 2, the middle needle on the 5, and the long needle on zero.

20.
0110. (Refer to figure 10.) Which altimeter(s) indicate(s) more than 10,000 feet?

A— A only.
B— B only.
C— A and B only.
D— A, B, and C.

Answer (C) is correct (0110). *(PHAK Chap III)*

Altimeters A and B indicate over 10,000 ft because A indicates 10,500 ft and B indicates 14,500 ft. The short needle on C points just below 1, i.e., below 10,000 ft.

FIGURE 10.—Altimeter.

Chapter 4: Airplane Instruments, Engines, and Systems

21.
0109. (Refer to figure 10.) Altimeter C indicates

A— 9,500 feet.
B— 10,950 feet.
C— 15,940 feet.
D— 19,500 feet.

Figure 10 is presented on page 59.

Answer (A) is correct (0109). *(PHAK Chap III)*
The altimeter in C is 9,500 ft because the shortest needle is near 1 (i.e., about 10,000 ft) the middle needle is between 9 and 0, indicating between 9,000 and 10,000 ft, and the long needle is on 5, indicating 500 ft.

Answer (B) is incorrect because, for 10,950 ft, the middle needle would have to be near the 1 and the long needle would have to be between 9 and 0. Answers (C) and (D) are incorrect because the short needle is near 1, i.e., close to 10,000 ft.

22.
0363. Prior to takeoff, the altimeter should be set to

A— the current local altimeter setting, if available, or the departure airport elevation.
B— the corrected density altitude of the departure airport.
C— the corrected pressure altitude for the departure airport.
D— 29.92" Hg.

Answer (A) is correct (0363). *(FAR 91.81)*
Prior to takeoff, the altimeter should be set to the local altimeter setting, or to the departure airport elevation.

Answer (B) is incorrect because density altitude is a measure of performance, not of actual altitude. Answers (C) and (D) are incorrect because pressure altitude is only used at or above 18,000 ft MSL.

23.
0362. If an altimeter setting is not available before flight, to which altitude or setting should the pilot adjust the altimeter?

A— To 29.92" Hg for flight below 18,000 feet MSL.
B— The elevation of the nearest airport corrected to mean sea level.
C— The elevation of the departure area.
D— Pressure altitude corrected for nonstandard temperature.

Answer (C) is correct (0362). *(FAR 91.81)*
When the local altimeter setting is not available at takeoff, the altimeter should be set to the elevation of the departure area.

Answers (A) and (D) are incorrect because pressure altitude is only used at or above 18,000 ft MSL. Answer (B) is incorrect because, if one corrects an elevation to mean sea level, it would only be correct at sea level.

TYPES OF ALTITUDE

24.
0002. Absolute altitude is the

A— altitude read directly from the altimeter.
B— altitude above the surface.
C— altitude reference to the standard datum plane.
D— indicated altitude corrected for instrument error.

Answer (B) is correct (0002). *(PHAK Chap III)*
Absolute altitude is altitude above the surface.

Answer (A) is incorrect because it is indicated altitude. Answer (C) is incorrect because it is pressure altitude. Answer (D) is incorrect because it is calibrated altitude.

25.
0001. What is true altitude?

A— Actual height above sea level corrected for all errors.
B— Altitude above the surface.
C— Altitude reference to the standard datum plane.
D— Altitude shown on a radar altimeter.

Answer (A) is correct (0001). *(PHAK Chap III)*
True altitude is the actual altitude above mean sea level.

Answers (B) and (D) are incorrect because each represents absolute altitude. Answer (C) is incorrect because it is pressure altitude.

Chapter 4: Airplane Instruments, Engines, and Systems

26.
0003. Density altitude is the

A— altitude reference to the standard datum plane.
B— pressure altitude corrected for nonstandard temperature.
C— altitude read directly from the altimeter.
D— altitude above the surface.

Answer (B) is correct (0003). *(AvW Chap 3)*

Density altitude is the pressure altitude corrected for nonstandard temperature.

Answer (A) is incorrect because it defines pressure altitude. Answer (C) is incorrect because it is indicated altitude. Answer (D) is incorrect because it is absolute altitude.

27.
0004. How can pressure altitude be determined?

A— Set the field elevation in the altimeter setting window and read the indicated altitude.
B— Set the altimeter to the field elevation and read the value in the altimeter setting window.
C— Set the altimeter to zero and read the value in the altimeter setting window.
D— Set 29.92 in the altimeter setting window and read the indicated altitude.

Answer (D) is correct (0004). *(FTH Chap 17)*

Pressure altitude is the indicated altitude when the altimeter is set at 29.92" Hg or 1013.2 millibars.

Answer (A) is incorrect because barometric pressure, not field elevation, is set in the altimeter setting window. Answer (B) is incorrect because it indicates the current barometric pressure, not standard barometric pressure. Answer (C) is a nonsense answer unless you are at sea level and you wish to determine the current altimeter setting.

28.
0014. If the altimeter indicates 1,380 feet when set to 29.92, what is the pressure altitude?

A— 1,280 feet.
B— 1,380 feet.
C— 1,480 feet.
D— 1,580 feet.

Answer (B) is correct (0014). *(AvW Chap 3)*

An altimeter indicates pressure altitude when set for 29.92. Accordingly, the altimeter indicating 1,380 ft when set for 29.92 indicates the pressure altitude is 1,380 ft.

29.
0010. Under what condition is pressure altitude and density altitude the same value?

A— At sea level, when the temperature is 0 °F.
B— When the altimeter has no installation error.
C— When the altimeter setting is 29.92.
D— At standard temperature.

Answer (D) is correct (0010). *(AvW Chap 3)*

Pressure altitude and density altitude are the same when temperature is standard.

Answer (A) is incorrect because standard temperature at sea level is 59°F. Answer (B) is incorrect because nonstandard temperature causes the difference between density and pressure altitude. Installation error relates to the airspeed indicator, not the altimeter. Answer (C) is incorrect because it describes pressure altitude.

30.
0011. Under what condition is indicated altitude the same as true altitude?

A— If the altimeter has no mechanical error.
B— When at sea level under standard conditions.
C— When at 18,000 feet with the altimeter set at 29.92.
D— At any altitude, if the indicated altitude is corrected for nonstandard sea level temperature and pressure.

Answer (B) is correct (0011). *(PHAK Chap III)*

Indicated altitude (what you read on your altimeter) approximates the true altitude (distance above mean sea level) when standard conditions exist and your altimeter is properly calibrated. Note you do not have to be at sea level for true altitude to equal indicated altitude.

Answers (A) and (C) are incorrect because the indicated altitude must be adjusted for nonstandard temperature and pressure. Answer (D) is incorrect because the indicated altitude must be corrected for nonstandard conditions existing where you are, not just as they exist at sea level.

Chapter 4: Airplane Instruments, Engines, and Systems

31.
0009. Under which condition(s) will pressure altitude be equal to true altitude?

A— When the atmospheric pressure is 29.92" Hg.
B— When standard atmospheric conditions exist.
C— When indicated altitude is equal to the pressure altitude.
D— When the outside air temperature (OAT) is standard for that altitude.

Answer (B) is correct (0009). *(FTH Chap 17)*
Pressure altitude equals true altitude when standard atmospheric conditions (29.92" Hg and 15°C at sea level) exist.
Answers (A) and (C) are incorrect because they ignore the possible differences from standard temperature. Answer (D) is incorrect because it ignores the barometric pressure differences.

32.
0005. Altimeter setting is the value to which the scale of the pressure altimeter is set so the altimeter indicates

A— density altitude at sea level.
B— pressure altitude at sea level.
C— true altitude at field elevation.
D— pressure altitude at field elevation.

Answer (C) is correct (0005). *(AvW Chap 3)*
Altimeter setting is the value to which the scale of the pressure altimeter is set so that the altimeter indicates true altitude at field elevation.
Answer (A) is incorrect because altimeters do not take into account temperature levels, which are required to determine density altitude. Answer (B) is incorrect because pressure altitude is determined by using the altimeter setting of 29.92. Answer (D) is incorrect because pressure altitude is the altitude above the surface, not above MSL.

ALTIMETER CALCULATIONS

33.
0035. If it is necessary to set the altimeter from 29.15 to 29.85, what change is made on the indicated altitude?

A— 70-foot increase.
B— 70-foot decrease.
C— 700-foot increase.
D— 700-foot decrease.

Answer (C) is correct (0035). *(AvW Chap 3)*
When increasing the altimeter setting from 29.15 to 29.85, the indicated altitude increases by 700 ft. The altimeter-indicated altitude moves in the same direction as the altimeter setting and changes about 1,000 ft for every change of 1" Hg in the altimeter setting.

34.
0019. If a pilot changes the altimeter setting from 30.11 to 29.96, the altimeter will indicate approximately

A— 15 feet higher.
B— 15 feet lower.
C— 150 feet higher.
D— 150 feet lower.

Answer (D) is correct (0019). *(AvW Chap 3)*
Atmospheric pressure decreases approximately 1" of mercury for every 1,000 ft of altitude gained. As an altimeter setting is changed, the change in altitude indication changes the same way (i.e., approximately 1,000 ft for every 1" change in altimeter setting) and in the same direction (i.e., lowering the altimeter setting lowers the altitude reading). Thus, changing from 30.11 to 29.96 is a decrease of .15 inches, or 150 ft (.15 x 1,000 ft) lower.

ALTIMETER ERRORS

35.
0015. If a flight is made from an area of low pressure into an area of high pressure without the altimeter setting being adjusted and a constant indicated altitude is maintained, the altimeter would indicate

A— the actual altitude above sea level.
B— the actual altitude above ground level.
C— higher than the actual altitude above sea level.
D— lower than the actual altitude above sea level.

Answer (D) is correct (0015). *(AvW Chap 3)*
When an altimeter setting is at a lower value than the actual local barometric pressure, the altimeter is indicating less than it should and thus, would be showing lower than the actual altitude above sea level.

Chapter 4: Airplane Instruments, Engines, and Systems

36.
0016. If a flight is made from an area of high pressure into an area of lower pressure without the altimeter setting being adjusted and a constant indicated altitude is maintained, the altimeter would indicate

A— the actual altitude above sea level.
B— the actual altitude above ground level.
C— higher than the actual altitude above sea level.
D— lower than the actual altitude above sea level.

Answer (C) is correct (0016). *(AvW Chap 3)*
When flying from higher pressure to lower pressure without adjusting your altimeter, you are losing true altitude if you maintain a constant indicated altitude. Thus, you indicate a higher than actual altitude.

37.
0013. Which condition would cause the altimeter to indicate a lower altitude than actually flown (true altitude)?

A— Air temperature lower than standard.
B— Atmospheric pressure lower than standard.
C— Pressure altitude the same as indicated altitude.
D— Air temperature warmer than standard.

Answer (D) is correct (0013). *(AvW Chap 3)*
In air that is warmer than standard temperature, the airplane will be higher than the altimeter indicates. Said another way, the altimeter will indicate a lower altitude than actually flown.

Answer (A) is incorrect because, when flying in air that is colder than standard temperature, the airplane will be lower than the altimeter indicates. Answer (B) is incorrect because, when atmospheric pressure is lower than standard, the altimeter indicates higher than actual altitude. Answer (C) is incorrect because when pressure altitude equals density altitude, temperature is standard and the altimeter should equal true altitude.

38.
0012. Under what condition will true altitude be lower than indicated altitude with an altimeter setting of 29.92 even with an accurate altimeter?

A— In colder than standard air temperature.
B— In warmer than standard air temperature.
C— When density altitude is higher than indicated altitude.
D— Under higher than standard pressure at standard air temperature.

Answer (A) is correct (0012). *(AvW Chap 3)*
The airplane will be lower than the altimeter indicates when flying in air that is colder than standard temperature. Remember that altimeter readings are adjusted for changes in barometric pressure but not for changes in temperature. When one flies from warmer to cold air and keeps a constant indicated altitude at a constant altimeter setting, the plane has actually descended.

Answer (B) is incorrect because the altimeter indicates lower than actual altitude in warmer than standard temperature. Answers (C) and (D) are incorrect because, when density altitude or actual pressure are higher than standard pressure, the altimeter indicates lower than actual altitude.

39.
0018. How do variations in temperature affect the altimeter?

A— Pressure levels are raised on warm days and the indicated altitude is lower than true altitude.
B— Higher temperatures expand the pressure levels and the indicated altitude is higher than true altitude.
C— Lower temperatures lower the pressure levels and the indicated altitude is lower than true altitude.
D— Indicated altitude varies directly with the temperature.

Answer (A) is correct (0018). *(AvW Chap 3)*
On warm days, the atmospheric pressure levels are higher than on cold days. Your altimeter will indicate a lower than true altitude.

Answers (B) and (C) are incorrect because they state the opposite of what is true. Remember that at colder than standard temperature, your true altitude will be below your indicated altitude. Answer (D) is incorrect because indicated altitude varies inversely with temperature.

ENGINE TEMPERATURE

40.
0075. Excessively high engine temperatures, either in the air or on the ground, will

A— cause damage to heat-conducting hoses and warping of the cylinder cooling fins.
B— cause loss of power, excessive oil consumption, and possible permanent internal engine damage.
C— increase fuel consumption and may increase power due to the increased heat.
D— not appreciably affect an aircraft engine in either environment.

Answer (B) is correct (0075). *(PHAK Chap II)*
Excessively high engine temperatures either in the air or on the ground will result in loss of power, excessive oil consumption, and possible permanent internal engine damage.
Answer (A) is incorrect because excessively high engine temperatures may cause internal engine damage, but external damage is less likely. Answer (C) is incorrect because excessive engine temperatures will not increase engine power. Answer (D) is incorrect because an excessively high engine temperature, either in the air or on the ground, can cause a loss of performance and possibly internal engine damage.

41.
0076. For internal cooling, reciprocating aircraft engines are especially dependent on

A— a properly functioning thermostat.
B— air flowing over the exhaust manifold.
C— the circulation of lubricating oil.
D— a lean fuel/air mixture.

Answer (C) is correct (0076). *(PHAK Chap II)*
An engine accomplishes some of its cooling by the flow of oil through the lubrication system. The lubrication system aids in cooling by reducing friction and absorbing heat from internal engine parts. Many airplane engines use an oil cooler, a small radiator device that will cool the oil before it is recirculated through the engine.
Answer (A) is incorrect because airplanes with air-cooled engines do not use thermostats. Answer (B) is incorrect because air flowing over the exhaust manifold would have little effect on cooling internal engine parts. Answer (D) is incorrect because a lean fuel/air mixture is likely to cause overheating.

42.
0074. An abnormally high engine oil temperature indication may be caused by

A— the oil level being too high.
B— the oil level being too low.
C— operating with a too high viscosity oil.
D— operating with an excessively rich mixture.

Answer (B) is correct (0074). *(PHAK Chap II)*
Operating with an excessively low oil level prevents the oil from being cooled adequately; i.e., an inadequate supply of oil will not be able to transfer engine heat to the engine's oil cooler (similar to a car engine's water radiator). Insufficient oil may also damage an engine from excessive friction within the cylinders and on other metal-to-metal contact parts.
Answer (A) is incorrect because an oversupply of oil results in lower engine oil temperature. Answer (C) is incorrect because the higher the viscosity, the better the lubricating and cooling capability of the oil. Answer (D) is incorrect because a rich fuel/air mixture does not increase engine oil temperature.

43.

0077. If the engine oil temperature and cylinder head temperature gauges have exceeded their normal operating range, the pilot may have been

A— operating with the mixture set too rich.
B— operating with higher-than-normal oil pressure.
C— operating with too much power and with the mixture set too lean.
D— using fuel that has a higher-than-specified fuel rating.

Answer (C) is correct (0077). *(PHAK Chap II)*

If the engine oil temperature and cylinder head temperature gauges exceed their normal operating range, it is possible that the power setting is too great and the fuel/air mixture is set excessively lean. These conditions may cause an increase in engine temperature.

Answer (A) is incorrect because a rich mixture setting does not normally cause higher-than-normal engine temperature. Answer (B) is incorrect because a higher-than-normal oil pressure does not normally increase the engine temperature. Answer (D) is incorrect because using fuel with a lower-, not higher-, than-specified rating might cause the engine to operate at a higher temperature.

44.

0246. What action can a pilot take to aid in cooling an engine that is overheating during a climb?

A— Lean the mixture to best power condition.
B— Increase RPM and reduce climb speed.
C— Reduce rate of climb and increase airspeed.
D— Increase RPM and climb speed.

Answer (C) is correct (0246). *(PHAK Chap II)*

If an airplane is overheating during a climb, the engine temperature will be decreased if the airspeed is increased. Airspeed will increase if the rate of climb is reduced.

Answers (A), (B), and (D) are incorrect because leaning the mixture and increasing RPM increase engine temperature.

45.

0247. What is one procedure to aid in cooling an engine that is overheating?

A— Enrichen the fuel mixture.
B— Increase the RPM.
C— Reduce the airspeed.
D— Use alternate air.

Answer (A) is correct (0247). *(PHAK Chap II)*

Increased (enriched) fuel mixtures have a cooling effect on an engine.

Answers (B) and (C) are incorrect because increasing the RPM and reducing airspeed increase engine temperature. Answer (D) is incorrect because the alternate air is not a cooling air; it is an alternate source of air for the carburetor.

CONSTANT-SPEED PROPELLER

46.

0227. How is engine operation controlled on an engine equipped with a constant-speed propeller?

A— The throttle controls power output as registered on the manifold pressure gauge and the propeller control regulates engine RPM.
B— The throttle controls power output as registered on the manifold pressure gauge and the propeller control regulates a constant blade angle.
C— The throttle controls engine RPM as registered on the tachometer and the mixture control regulates the power output.
D— The throttle controls engine RPM as registered on the tachometer and the propeller control regulates the power output.

Answer (A) is correct (0227). *(PHAK Chap III)*

Airplanes equipped with controllable pitch propellers have both a throttle control and a propeller control. The throttle controls the power output of the engine, which is registered on the manifold pressure gauge. This is a simple barometer that measures the air pressure in the engine intake manifold in inches of mercury. The propeller control regulates the engine RPM, which is registered on a tachometer.

Answer (B) is incorrect because the blade angle changes to control the RPM. Answers (C) and (D) are incorrect because the throttle controls power output, and the propeller control regulates the RPM.

47.
0228. A precaution for the operation of an engine equipped with a constant-speed propeller is to

A— avoid high RPM settings with high manifold pressure.
B— avoid high RPM settings with low manifold pressure.
C— always use a rich mixture with high RPM settings.
D— avoid high manifold pressure settings with low RPM.

Answer (D) is correct (0228). *(PHAK Chap III)*
For any given RPM, there is a manifold pressure that should not be exceeded. Manifold pressure is excessive for a given RPM when the pressure that the cylinders were designed for can be exceeded, placing undue stress on them. If repeated or extended, the stress would weaken the cylinder components and eventually cause engine failure.

Answers (A) and (B) are incorrect because it is the relationship of high manifold pressure with low RPM that is dangerous. Answer (C) is incorrect because the mixture control is related to engine cylinder temperature, not to manifold pressure or RPM.

48.
0229. What is an advantage of a constant-speed propeller?

A— Permits the pilot to select and maintain a desired cruising speed.
B— Allows a higher cruising speed than possible with a fixed-pitch propeller.
C— Provides a smoother operation with stable RPM and eliminates vibrations.
D— Permits the pilot to select the blade angle for the most efficient performance.

Answer (D) is correct (0229). *(PHAK Chap III)*
A controllable-pitch propeller (constant-speed) permits the pilot to select the blade angle that will result in the most efficient performance given the flight conditions. A low blade angle and a decreased pitch reduces the propeller drag and allows more engine power for takeoffs. After airspeed is attained during cruising flight, the propeller blade is changed to a higher angle to increase pitch. The blade takes a larger bite of air and a lower power setting and consequently increases the efficiency of the flight. This process is similar to shifting gears in an automobile from low to high.

ENGINE IGNITION SYSTEMS

49.
0073. One purpose of the dual ignition system on an aircraft engine is to provide for

A— improved engine performance.
B— uniform heat distribution.
C— balanced cylinder head pressure.
D— easier starting.

Answer (A) is correct (0073). *(PHAK Chap II)*
Most airplane engines are equipped with dual ignition systems, which have two magnetos to supply the electrical current to two spark plugs for each combustion chamber. The main advantages of the dual system are increased safety and improved burning and combustion of the mixture, which results in improved performance.

Answer (B) is incorrect because the heat distribution within a cylinder is usually not uniform. Answer (C) is incorrect because balanced cylinder-head pressure is a nonsense phrase. Answer (D) is incorrect because the primary purposes are safety and improved performance.

Chapter 4: Airplane Instruments, Engines, and Systems

50.
0072. If an engine continues to run after the ignition switch is turned to the OFF position, the probable cause may be

A— the mixture is too lean and this causes the engine to diesel.
B— the voltage regulator points are sticking closed.
C— a broken magneto ground wire.
D— fouled spark plugs.

Answer (C) is correct (0072). *(PHAK Chap II)*
If an airplane continues to run after the ignition switch is turned off, the cause is usually a broken or loosely-connected magneto ground wire. Lacking a ground, the magneto will continue its electrical output, and the spark plugs will continue to fire, resulting in combustion in the cylinders.

Answer (A) is incorrect because an engine diesels (runs on) as a result of fuel being ignited by the cylinder's heat after the spark plug ignition is stopped. The way to prevent dieseling is to stop the fuel flow. Answer (B) is incorrect because the generator-voltage regulator system is completely independent from the magneto-spark plug system. Answer (D) is incorrect because fouled spark plugs will not fire, regardless of whether the ignition switch is off or on.

CARBURETOR ICING

51.
0065. In comparison to fuel injection systems, float-type carburetor systems are generally considered to be

A— equally susceptible to icing as a fuel injection unit.
B— susceptible to icing only when visible moisture is present.
C— more susceptible to icing than a fuel injection unit.
D— less susceptible to icing than a fuel injection unit.

Answer (C) is correct (0065). *(PHAK Chap II)*
Float-type carburetor systems are generally more susceptible to icing than fuel-injected engines.

When there is moisture in the air (not necessarily visible) and the carburetor air is at the proper temperature, i.e., below freezing at the venturi tube, icing will occur.

52.
0066. The operating principle of float-type carburetors is based on the

A— automatic metering of air at the venturi as the aircraft gains altitude.
B— difference in air pressure at the venturi throat and the air inlet.
C— increase in air velocity in the throat of a venturi causing an increase in air pressure.
D— measurement of the fuel flow into the induction system.

Answer (B) is correct (0066). *(PHAK Chap II)*
In a float-type carburetor, air flows into the carburetor and through a venturi tube (a narrow throat in the carburetor). As the air flows rapidly through the venturi, a low pressure area is created which draws the fuel from a main fuel jet located at the throat of the carburetor and into the airstream, where it is mixed with flowing air. It is called a float-type carburetor in that a ready supply of gasoline is kept in the float bowl by a float, which activates a fuel inlet valve.

Answer (A) is incorrect because the metering of fuel, not air, at the venturi is done manually with a mixture control. Answer (C) is incorrect because the increase in air velocity in the throat of a venturi causes a decrease in air pressure (which draws the gas from the main fuel jet into the low-pressure air). Answer (D) is incorrect because fuel-injection systems measure the fuel flow into the engine.

53.
0070. If an aircraft is equipped with a fixed-pitch propeller and a float-type carburetor, the first indication of carburetor ice would most likely be

A— a drop in oil temperature and cylinder head temperature.
B— engine roughness.
C— a drop in manifold pressure.
D— loss of RPM.

Answer (D) is correct (0070). *(PHAK Chap II)*
In an airplane equipped with a fixed-pitch propeller and float-type carburetor, the first indication of carburetor ice would be a loss in RPM.
Answer (A) is incorrect because a carburetor icing condition does not cause a drop in oil temperature or cylinder head temperature. Answer (B) is incorrect because a loss in engine RPM would be evident before engine roughness became noticeable. Answer (C) is incorrect because a fixed-pitch propeller airplane does not usually have a manifold pressure gauge.

54.
0067. The presence of carburetor ice in an aircraft equipped with a fixed-pitch propeller can be verified by applying carburetor heat and noting

A— an increase in RPM and then a gradual decrease in RPM.
B— a decrease in RPM and then a gradual increase in RPM.
C— a decrease in RPM and then a constant RPM indication.
D— an immediate increase in RPM with no further change in RPM.

Answer (B) is correct (0067). *(PHAK Chap II)*
The presence of carburetor ice in an airplane equipped with a fixed-pitch propeller can be verified by applying carburetor heat and noting a decrease in RPM and then a gradual increase. The decrease in RPM as heat is applied is caused by less dense hot air entering the engine and reducing power output. Also, if ice is present, melting water entering the engine may also cause a loss in performance. As the carburetor ice melts, however, the RPM gradually increases until it stabilizes when the ice is completely removed.

55.
0441. Which conditions are favorable to the development of carburetor icing?

A— Any temperature below freezing, and a relative humidity of less than 50 percent.
B— Temperature between 32 °F and 50 °F, and low humidity.
C— Temperature between 0 °F and 20 °F, and high humidity.
D— Temperature between 20 °F and 70 °F, and high humidity.

Answer (D) is correct (0441). *(PHAK Chap II)*
When the temperature is between 20°F and 70°F with visible moisture or high humidity, one should be on the alert for carburetor ice. During low or closed throttle settings, an engine is particularly susceptible to carburetor icing.
Answers (A) and (B) are incorrect because, when relative humidity is low, icing is generally not a problem. Answer (C) is incorrect because icing is generally not a problem below 20°F.

56.
0442. The possibility of carburetor icing should always be considered when operating in conditions where the

A— temperature is as high as 95 °F, and the relative humidity is 30 percent or greater.
B— relative humidity range is from 25 percent to 100 percent, regardless of temperature.
C— relative humidity is between 30 percent and 100 percent, and the temperature is between 0 °F and 32 °F.
D— temperature is as high as 70 °F, and the relative humidity is greater than 50 percent.

Answer (D) is correct (0442). *(PHAK Chap II)*
When the temperature is between 20°F and 70°F with visible moisture or high humidity, one should be on the alert for carburetor ice. During low or closed throttle settings, an engine is particularly susceptible to carburetor icing.
Answers (A), (B), and (C) are incorrect because icing is generally not a problem when the temperature is below 20°F or above 70°F.

Chapter 4: Airplane Instruments, Engines, and Systems

CARBURETOR HEAT

57.
0071. The use of carburetor heat tends to

A— increase engine output and increase operating temperature.
B— decrease engine output and increase operating temperature.
C— increase engine output and decrease operating temperature.
D— decrease engine output and decrease operating temperature.

Answer (B) is correct (0071). *(PHAK Chap II)*
Use of carburetor heat reduces the engine output and increases the operating temperature. Thus, carburetor heat should not be used when full power is required (as during takeoff) or during normal engine operation except as a check for the presence or removal of carburetor ice.
Answers (A) and (C) are incorrect because carburetor heat decreases (not increases) engine output. Answers (C) and (D) are incorrect because carburetor heat increases (not decreases) operating temperature.

58.
0068. Applying carburetor heat will

A— result in more air going through the carburetor.
B— enrich the fuel/air mixture.
C— lean the fuel/air mixture.
D— not affect the mixture.

Answer (B) is correct (0068). *(PHAK Chap II)*
Applying carburetor heat will enrich the fuel/air mixture. Warm air is less dense than cold air, hence the application of heat increases the fuel-to-air ratio.
Answer (A) is incorrect because applying carburetor heat will not result in more air going into the carburetor. Answers (C) and (D) are incorrect because applying carburetor heat will enrich the fuel/air mixture.

59.
0069. What change occurs in the fuel/air mixture when carburetor heat is applied?

A— A decrease in RPM results from the lean mixture.
B— The fuel/air mixture becomes leaner.
C— The fuel/air mixture becomes richer.
D— No change occurs in the fuel/air mixture.

Answer (C) is correct (0069). *(PHAK Chap II)*
When carburetor heat is applied, hot air is introduced into the carburetor. Hot air is less dense than cold air; therefore, the decrease in air density with a constant amount of fuel makes a richer mixture.
Answer (A) is incorrect because a drop in RPM as carburetor heat is applied is due to the less dense air and melting ice, not a lean mixture. Answers (B) and (D) are incorrect because, when carburetor heat is applied, the fuel/air mixture becomes richer, not leaner.

FUEL/AIR MIXTURE

60.
0088. During the runup at a high-elevation airport, a pilot notes a slight engine roughness that is not affected by the magneto check but grows worse during the carburetor heat check. Under these circumstances, which would be the most logical initial action?

A— Check the results obtained with a leaner setting of the mixture control.
B— Taxi back to the flight line for a maintenance check.
C— Reduce manifold pressure to control detonation.
D— Check to see that the mixture control is in the full-rich position.

Answer (A) is correct (0088). *(PHAK Chap II)*
If, during a run-up at a high-elevation airport, you notice a slight roughness that is not affected by a magneto check but grows worse during the carburetor heat check, you should check the results obtained with a leaner setting of the mixture control. At a high-elevation field, the air is less dense and the application of carburetor heat increases the already too rich fuel-to-air mixture. By leaning the mixture during the run-up, the condition should improve.
Answer (B) is incorrect because this mixture condition is normal at a high-elevation field. However, if after leaning the mixture a satisfactory run-up cannot be obtained, the pilot should taxi back to the flight line for a maintenance check. Answer (C) is incorrect because the question describes a symptom of an excessively rich mixture, not detonation. Answer (D) is incorrect because the problem arises from too rich a mixture.

61.
0086. The basic purpose of adjusting the fuel/air mixture control at altitude is to

A— decrease the amount of fuel in the mixture in order to compensate for increased air density.
B— decrease the fuel flow in order to compensate for decreased air density.
C— increase the amount of fuel in the mixture to compensate for the decrease in pressure and density of the air.
D— increase the fuel/air ratio for flying at altitude.

Answer (B) is correct (0086). *(PHAK Chap II)*
The purpose of adjusting the fuel/air mixture control at high altitudes is to decrease the fuel flow to compensate for the decreased air density.
Answer (A) is incorrect because air density decreases, not increases, at altitude. Answer (C) is incorrect because, the mixture is decreased (not increased) in order to compensate for decreased air density. Answer (D) is incorrect because, due to less dense air at high altitudes, the mixture must be leaned (not increased or enriched) for best performance.

62.
0087. While cruising at 9,500 feet MSL, the fuel/air mixture is properly adjusted. If a descent to 4,500 feet MSL is made without readjusting the mixture control,

A— the fuel/air mixture may become excessively rich.
B— the fuel/air mixture may become excessively lean.
C— there will be more fuel in the cylinders than is needed for normal combustion, and the excess fuel will absorb heat and cool the engine.
D— the excessively rich mixture will create higher cylinder head temperatures and may cause detonation.

Answer (B) is correct (0087). *(PHAK Chap II)*
At 9,500 ft, the mixture adjustment will be lean in order for the engine to run properly. As the airplane descends, the density of the air increases and there will be less fuel to air in the ratio, causing a leaner running engine. This excessively lean mixture will create higher cylinder temperature and may cause detonation.
Answers (A), (C), and (D) are incorrect because the fuel/air mixture will be leaner, not richer.

ABNORMAL COMBUSTION

63.
0079. Detonation occurs in a reciprocating aircraft engine when

A— the spark plugs are fouled or shorted out or the wiring is defective.
B— hot spots in the combustion chamber ignite the fuel/air mixture in advance of normal ignition.
C— the fuel/air mixture is too rich.
D— the unburned charge in the cylinders explodes instead of burning normally.

Answer (D) is correct (0079). *(FTH Chap 2)*
Detonation occurs when the fuel/air mixture in the cylinders explodes instead of burning normally. This more rapid force slams the piston down instead of pushing it.
Answer (A) is incorrect because, if the spark plugs are "fouled" or the wiring is defective, the cylinders would not be firing; i.e., there would be no combustion. Answer (B) is incorrect because hot spots in the combustion chamber igniting the fuel/air mixture in advance of normal ignition is preignition. Answer (C) is incorrect because too rich a fuel/air mixture, resulting in cooler cylinder head temperatures, deters (not causes) both detonation and preignition.

Chapter 4: Airplane Instruments, Engines, and Systems

64.
0080. If a pilot suspects that the engine (with a fixed-pitch propeller) is detonating during climb-out after takeoff, normally the corrective action to take would be to

A— increase the rate of climb.
B— retard the throttle.
C— lean the mixture.
D— apply carburetor heat.

Answer (B) is correct (0080). *(FTH Chap 2)*
If you suspect engine detonation during climb-out after takeoff, you would normally retard the throttle slightly, thus decreasing the engine RPM. Detonation is usually caused by a poor grade of fuel or an excessive engine temperature.
Answer (A) is incorrect because increasing the rate of climb by using a steeper climb angle will put a greater load on the engine, increase engine temperature, and increase detonation. Answer (C) is incorrect because leaning the mixture will increase engine temperature and increase detonation. Answer (D) is incorrect because, while carburetor heat will increase the fuel-to-air ratio, hot air flowing into the carburetor will not lower engine temperature. Also, the less dense air will decrease the engine power for climb-out.

65.
0081. If the grade of fuel used in an aircraft engine is lower than specified for the engine, it will most likely cause

A— a mixture of fuel and air that is not uniform in all cylinders.
B— lower cylinder head temperatures.
C— an increase in power which could overstress internal engine components.
D— detonation.

Answer (D) is correct (0081). *(FTH Chap 2)*
If the grade of fuel used in an airplane engine is lower than specified for the engine, it will probably cause detonation. Lower grades of fuel ignite at lower temperatures. A higher temperature engine (which should use a higher grade of fuel) may cause lower grade fuel to explode (detonate) rather than burn evenly.
Answer (A) is incorrect because the carburetor meters the lower-grade fuel quantity in the same manner as a higher grade of fuel. Answer (B) is incorrect because a lower grade of fuel will cause higher cylinder head temperatures. Answer (C) is incorrect because the use of fuel of a grade lower than specified will likely cause a decrease in power.

66.
0082. The uncontrolled firing of the fuel/air charge in advance of normal spark ignition is known as

A— combustion.
B— pre-ignition.
C— atomizing.
D— detonation.

Answer (B) is correct (0082). *(FTH Chap 2)*
Preignition is the ignition of the fuel prior to normal ignition or ignition before the electrical arcing occurs at the spark plug. Preignition may be caused by excessively hot exhaust valves, carbon particles, or spark plugs and electrodes heated to an incandescent, or glowing, state. These hot spots are usually caused by high temperatures encountered during detonation. A significant difference between preignition and detonation is that if the conditions for detonation exist in one cylinder they may exist in all cylinders, but preignition may take place in only one or two cylinders.

Chapter 4: Airplane Instruments, Engines, and Systems

AVIATION FUEL PRACTICES

67.
0085. What type fuel can be substituted for an aircraft if the recommended octane is not available?

A— The next higher octane aviation gas.
B— The next lower octane aviation gas.
C— Unleaded automotive gas of the next higher rating.
D— Unleaded automotive gas of the same octane rating.

Answer (A) is correct (0085). *(PHAK Chap II)*

If the recommended octane is not available for an airplane, the next higher octane aviation gas should be used.

Answer (B) is incorrect because if the grade of fuel used in an airplane engine is lower than specified for the engine, it will probably cause detonation. Answers (C) and (D) are incorrect because, except for very special situations, only aviation gas should be used.

68.
0078. Filling the fuel tanks after the last flight of the day is considered a good operating procedure because this will

A— force any existing water to the top of the tank away from the fuel lines to the engine.
B— prevent expansion of the fuel by eliminating airspace in the tanks.
C— prevent moisture condensation by eliminating airspace in the tanks.
D— eliminate vaporization of the fuel.

Answer (C) is correct (0078). *(PHAK Chap II)*

Filling the fuel tanks after the last flight of the day is considered good operating practice because it prevents moisture condensation by eliminating airspace in the tanks.

Answer (A) is incorrect because water is heavier than fuel and will always settle to the bottom of the tank. Answers (B) and (D) are incorrect because filling the fuel tank will prevent neither expansion nor vaporization of the fuel.

69.
0083. On aircraft equipped with fuel pumps, the practice of running a fuel tank dry before switching tanks is considered unwise because

A— the engine-driven fuel pump or electric fuel boost pump may draw air into the fuel system and cause vapor lock.
B— the engine-driven fuel pump is lubricated by fuel and operating on a dry tank may cause pump failure.
C— any foreign matter in the tank will be pumped into the fuel system.
D— the fuel pump is located above the bottom portion of the fuel tank.

Answer (A) is correct (0083). *(PHAK Chap II)*

Running one gas tank dry in flight before switching to another tank is dangerous. The fuel pump(s) may draw air into the fuel system and cause "vapor lock," which prevents the fuel pump from pumping gas. A vapor-locked engine may be extremely difficult to restart.

Answer (B) is incorrect because the engine does not operate on a dry fuel tank. Answer (C) is incorrect because foreign matter in the bottom of gas tanks is usually drained from the fuel system prior to flight. Answer (D) is incorrect because, no matter where the fuel pump is located, air will be pumped into the fuel lines when the fuel level is as low as the line out of which fuel flows from the tank.

70.
0084. Which would most likely cause the cylinder head temperature and engine oil temperature gauges to exceed normal operating range?

A— Using fuel that has a higher-than-specified fuel rating.
B— Using fuel that has a lower-than-specified fuel rating.
C— Operating with higher-than-normal oil pressure.
D— Operating with the mixture control set too rich.

Answer (B) is correct (0084). *(PHAK Chap II)*

Use of fuel with lower-than-specified fuel ratings, e.g., 80 octane instead of 100, can cause many problems, including higher operating temperatures, detonation, etc.

Answers (A) and (D) are incorrect because both rich mixtures and higher octane fuels result in lower cylinder head temperatures. Answer (C) is incorrect because higher-than-normal oil pressure provides better lubrication and cooling (although too high an oil pressure can break parts, lines, etc.).

CHAPTER FIVE
AIRPORTS AND AIR TRAFFIC CONTROL

Runway Markings	(6 questions)	. . . . 73, 77
Airport Traffic Patterns	(8 questions)	. . . . 74, 79
Beacons	(1 question)	. . . . 75, 81
UNICOM and MULTICOM	(2 questions)	. . . . 75, 82
Visual Approach Slope Indicators (VASIs)	(6 questions)	. . . . 75, 82
Airport Traffic Areas	(5 questions)	. . . . 76, 84
Airport Advisory Areas	(2 questions)	. . . . 76, 85
TCAs and Airport Radar Service Areas	(4 questions)	. . . . 76, 85

This chapter contains outlines of major concepts tested, all FAA test questions and answers regarding airports and Air Traffic Control, and an explanation of each answer. The subtopics or modules within this chapter are listed above, followed in parentheses by the number of questions from the FAA written test pertaining to that particular module. The two numbers following the parentheses are the page numbers on which the outline and questions begin for that module.

CAUTION: Recall that the sole purpose of this book is to expedite your passing the FAA written test for the recreational pilot certificate. Accordingly, all extraneous material (i.e., topics or regulations not directly tested on the FAA written test) is omitted, even though much more information and knowledge are necessary to fly safely. This additional material is presented in *RECREATIONAL PILOT FLIGHT MANEUVERS* and *PRIVATE PILOT HANDBOOK,* available from Aviation Publications, Inc. See pages 208 to 210 for more information and an order form.

RUNWAY MARKINGS (6 questions)

1. The number at the end of each runway indicates the magnetic alignment divided by 10°; e.g., runway 26 indicates 260° magnetic, runway 9 indicates 90° magnetic.

2. A **displaced threshold** (unsuitable for landing) is indicated by arrows from the end of a runway to a broad solid line across the runway. The remainder of the runway, following the displaced threshold, is the landing portion of the runway.

    a. The paved area behind the displaced threshold is available for taxiing, landing rollout, and takeoff.

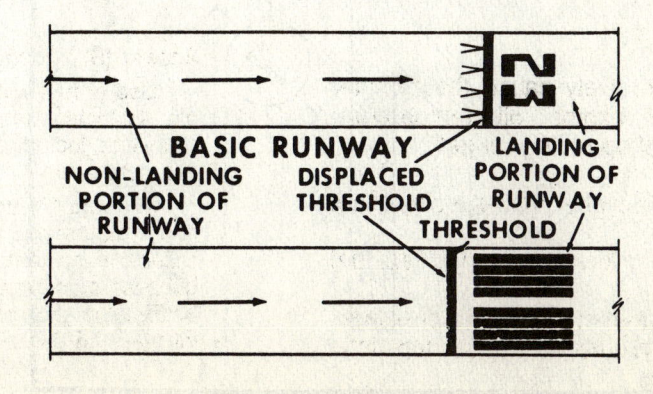

b. If the displaced threshold is marked by "chevrons," it is not available for any use, i.e., not even taxiing.

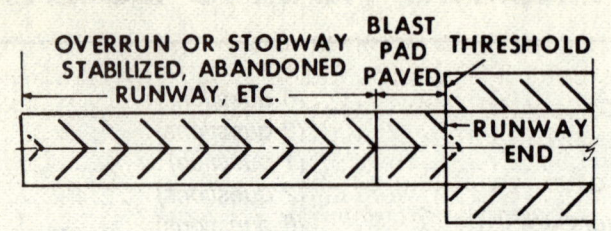

3. **Closed runways** are marked by an "X" on each runway end that is closed.

AIRPORT TRAFFIC PATTERNS (8 questions)

1. You land and takeoff into the wind to reduce your groundspeed (requires less runway and avoids excessive wear and tear).

    a. Runway numbers indicate magnetic direction after you add a zero, e.g., Runway 6 means 60°.

2. The segmented circle system provides traffic pattern information at airports without operating control towers. It consists of the

    a. *Segmented circle* -- located in a position affording maximum visibility to pilots in the air and on the ground, and providing a centralized point for the other elements of the system.

    b. *Landing strip indicators* -- angular indicators (legs sticking out of the segmented circle) showing the direction of flight to and from a runway.

    1) In the example below, runways 22 and 36 are left hand, while runways 4 and 18 are right hand.
    2) The "X" indicates a runway is closed.

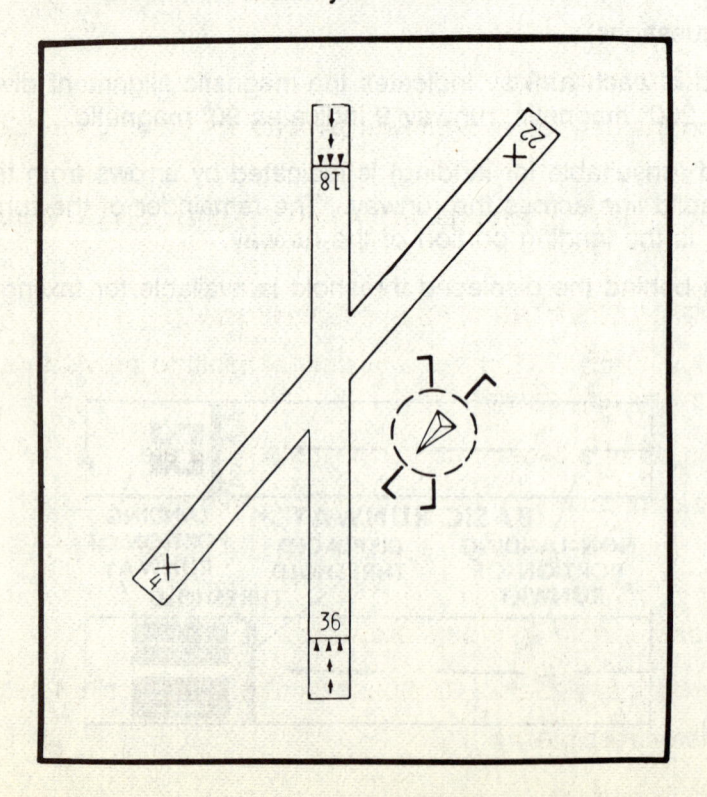

Chapter 5: Airports and Air Traffic Control 75

- c. *Wind direction indicator* -- A wind cone, wind sock, or wind "T" is installed near the runways to indicate wind direction.
- d. *Landing direction indicator* -- A tetrahedron on a swivel is installed when conditions at the airport warrant its use. It is used to indicate the direction of takeoffs and landings. It should be located at the center of a segmented circle and may be lighted for night operations.
    1) The small end points toward the direction in which a takeoff or landing should be made; i.e., the small end points into the wind.

3. If there is no segmented circle installed at the airport, traffic pattern indicators may be installed on or near the end of the runway.

4. Remember, you land
    a. In the same direction as the tip of a tetrahedron is pointing,
    b. As if you were flying out of the large (open) end of the wind cone, or
    c. Toward the cross-bar end of a wind "T" (visualize the "T" as an airplane with no nose, with the top of the "T" being the wings).

5. If you are approaching an airport without an operating control tower,
    a. You must turn to the left when landing unless visual displays advise otherwise.
    b. You must comply with any FAA traffic pattern for that airport when departing.

BEACONS (1 question)

1. Operation of the green and white rotating beacon in a control zone during the day indicates that the weather is not VFR, i.e.,
    a. The visibility is less than three miles, or
    b. The ceiling is less than 1,000 ft.

UNICOM AND MULTICOM (2 questions)

1. UNICOM communication frequencies at nontower airports are commonly 122.7, 122.8, and 123.0.

2. MULTICOM is 122.9 and is used at airports without an assigned UNICOM frequency.

VISUAL APPROACH SLOPE INDICATORS (VASIs) (6 questions)

1. *Visual approach slope indicators (VASIs)* are a system of lights to provide visual descent information during runway approaches.

2. The standard VASI consists of a two-barred tier of lights. You are
    a. Below the glidepath if both light bars are red, i.e., "red means dead."
    b. On the glidepath if the far (on top visually) lights are red and the near (on bottom visually) lights are white.
    c. Above the glidepath if both light bars are white.

3. Remember, red over white (i.e., R before W alphabetically) is the desired sequence.
    a. White over red is impossible.

4. A tri-color VASI is a single light unit projecting three colors.
    a. The below glidepath indicator is red.
    b. The above glidepath indicator is amber.
    c. The on glidepath indicator is green.
5. VASI only determines the proper height. It has no bearing on runway alignment.
6. FARs require you to maintain an altitude at or above VASI glide slopes.

AIRPORT TRAFFIC AREAS (5 questions)

1. Airport traffic areas exist at each airport with an operating control tower
    a. To control airplanes operating to, from, or on airports with operating control towers.
    b. When a control tower is not operating, the airport traffic area is not in effect.
2. *An airport traffic area* is the airspace
    a. Within a horizontal radius of 5 statute miles from the geographical center of any airport with an operating control tower, AND
    b. Up to, but not including, an altitude of 3,000 ft above the surface.
3. Two-way radio communication with the control tower is required for landings and takeoffs in an airport traffic area.

AIRPORT ADVISORY AREAS (2 questions)

1. Airport advisory areas exist at noncontrolled airports that have a Flight Service Station (FSS).
    a. The FSS provides advisory (not control) information on traffic, weather, etc. to requesting aircraft, i.e., both arriving and departing aircraft.

TCAS AND AIRPORT RADAR SERVICE AREAS (4 questions)

1. *Terminal Control Areas (TCAs)* are controlled airspace usually found around larger airports with high volumes of traffic.
2. *Airport Radar Service Area (ARSA)* is a special type of controlled airspace which REQUIRES communication with ATC.
    a. Consists of an inner circle and outer circle and an outer area.
        1) The inner circle is a 5 nautical mile (NM) radius from primary airport.
            a) Surface to 4,000 ft AGL.
        2) The outer circle is a 5-10 NM band extending outward from the inner circle.
            a) 1,200 ft AGL to 4,000 ft AGL.
        3) The outer area extends outward from the primary airport with about a 20 NM radius.
            a) Extends from the lower limits of the radar coverage up to the ceiling of approach control's delegated airspace.
3. ARSAs and TCAs are depicted on sectional charts and terminal area charts.
4. Unless directed otherwise, the VFR transponder code is 1200.

Chapter 5: Airports and Air Traffic Control

> **QUESTIONS AND ANSWER EXPLANATIONS**
>
> All of the FAA questions from the written test for the Recreational Pilot certificate relating to airports and air traffic control and the material outlined previously are reproduced below in the same modules as the previous outlines. To the immediate right of each question are the correct answer and answer explanation. You should cover these answers and answer explanations while responding to the questions. Refer to the general discussion in Chapter 1 on how to take the examination.
>
> Remember that the questions from the FAA's Written Test Book have been reordered by topic, and the topics have been organized into a meaningful sequence. Accordingly, the first line of the answer explanation gives the FAA question number and the citation of the authoritative source for the answer.

RUNWAY MARKINGS

1.
0269. The numbers 9 and 27 on a runway indicate that the runway is oriented approximately

A— 009° and 027° magnetic.
B— 009° and 027° true.
C— 090° and 270° magnetic
D— 090° and 270° true.

Answer (C) is correct (0269). *(AIM Para 226)*
When numbering a runway, the closest 10° magnetic heading is used, and the last digit, which is always a zero, is dropped. These numbers are always magnetic and are determined by using a compass.

2.
0270. How is a runway recognized as being closed?

A— The letter C is painted in red after the runway number.
B— Red lights are placed at the approach end of the runway.
C— Yellow chevrons are painted on the runway beyond the threshold.
D— "X" is displayed on the runway.

Answer (D) is correct (0270). *(AIM Para 60)*
Closed runways are marked with an "X" displayed on each runway end that is closed, with the "X" appearing just above the runway number, if any.
Answer (A) is incorrect because the letter "C" indicates the central runway of three parallel runways. Answer (B) is incorrect because red lights do not indicate the runway is closed. They indicate the departure end of a runway in use. Answer (C) is incorrect because chevron markings mean that a portion of the runway is unsafe and should not be used for taxiing, takeoff, or landing (i.e., may be used only for emergency overrun).

3.
0263. (Refer to figure 27.) According to the airport diagram,

A— takeoffs and landings are permissible at position C since this is a short takeoff and landing runway.
B— the takeoff and landing portion of Rwy 12 begins at position B.
C— Rwy 30 is equipped at position E with emergency arresting gear to provide a means of stopping military aircraft.
D— takeoffs may be started at position A on Rwy 12, and the landing portion of this runway begins at position B.

Answer (D) is correct (0263). *(AIM Para 60)*
In Figure 27 below, Runway 12 takeoffs may be started at Position A, and the landing portion of this runway begins at Position B. In this example, a displaced threshold exists at the beginning of Runway 12. The threshold is a heavy line across the runway, designating the beginning portion of a runway usable for landing. Since the area short of the displaced threshold in Runway 12 does not have the chevron markings, an airplane may use this portion of the runway for takeoff.
Answer (A) is incorrect because Position C is on a closed runway (indicated by the "X"s). Answer (B) is incorrect because one may takeoff on the displaced threshold because it does not have chevron markings. Answer (C) is incorrect because arresting cables are indicated by yellow circles painted across the runway where the cables are located. Runway 30 does not contain these markings.

Figure 27 is presented on the following page.

Chapter 5: Airports and Air Traffic Control

4.
0264. (Refer to figure 27.) What is the difference between area A and area E on the airport?

A— "A" may be used for taxi and takeoff; "E" may be used only as an overrun.
B— "A" may be used for all operations except heavy aircraft landings; "E" may be used only as an overrun.
C— "A" may be used only for taxiing; "E" may be used for all operations except landings.
D— "A" may be used only for an overrun; "E" may be used only for taxi and takeoff.

Answer (A) is correct (0264). *(AIM Para 60)*
Area A may be used for taxi and takeoff. Area E may only be used for overrun.

Answer (B) is incorrect because the displaced theshold at A may only be used for taxi and takeoff, i.e., not for landings. Answers (C) and (D) are incorrect because chevron-marked displaced thresholds may be used for emergency overruns only.

5.
0265. (Refer to figure 27.) Area C on the airport is classified as

A— an STOL runway.
B— a parking ramp.
C— a multiple heliport.
D— a closed runway.

Answer (D) is correct (0265). *(AIM Para 60)*
The runway marked by the arrow C has "X"s on it, indicating it is closed.

Answer (A) is incorrect because STOL runways have "STOL" in runway-number-size letters printed on the end of the runway. Answer (B) is incorrect because parking ramps are paved areas away from runways and taxiways. Answer (C) is incorrect because heliports are marked by "H," not "X."

6.
0262. (Refer to figure 27.) That portion of the runway identified by the letter A

A— may be used for taxiing but should not be used for takeoffs or landings.
B— may be used for taxiing or takeoffs but not for landings.
C— may be used for taxiing, takeoffs, and landings.
D— may not be used except in an emergency.

Answer (B) is correct (0262). *(AIM Para 60)*
The portion of the runway identified by the letter A in Fig. 27 is a displaced threshold, which means it may be used for taxiing or takeoffs, but not for landings.

Answer (A) is incorrect because takeoffs are permitted as well as taxiing. Answer (C) is incorrect because landings are not permitted on displaced thresholds. Answer (D) is incorrect because only displaced thresholds with chevron markings are precluded from use except in emergencies.

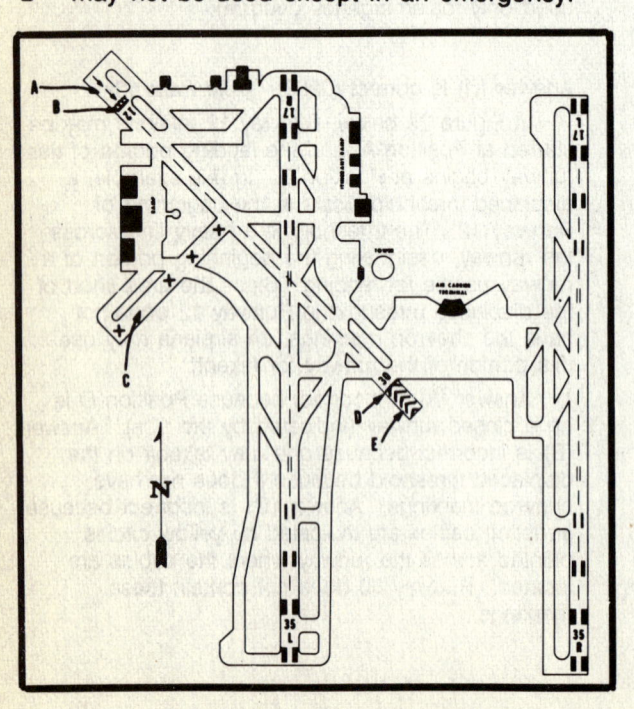

FIGURE 27.—Airport Diagram.

80   Chapter 5: Airports and Air Traffic Control

Figure 25 is presented on page 79.

10.
0253. (Refer to figure 25.) Which runway(s) and traffic pattern(s) should be used as indicated by the wind cone in the segmented circle?

A— Right-hand traffic on Rwy 35.
B— Right-hand traffic on Rwy 17.
C— Left-hand traffic on Rwy 35 or right-hand traffic on Rwy 26.
D— Left-hand traffic on Rwy 26 or Rwy 35.

Answer (C) is correct (0253). *(AIM Para 222)*
The appropriate landing, given a wind from the northwest (Fig. 25), is a left-hand on Runway 35 or a right-hand on Runway 26. Each would have a quartering headwind.

Answer (A) is incorrect because Runway 35 uses a left-hand pattern, also this would be a tail-wind landing. Answer (B) is incorrect because, even though there is right traffic on Runway 17, you would be landing with a tailwind. Answer (D) is incorrect because there is right-hand, not left-hand, traffic on Runway 26.

11.
0267. (Refer to figure 28.) If the wind is as shown by the landing direction indicator, the pilot should land to the

A— north on Rwy 36 and expect a crosswind from the right.
B— south on Rwy 18 and expect a crosswind from the right.
C— southwest on Rwy 22 directly into the wind.
D— northeast on Rwy 4 directly into the wind.

Answer (B) is correct (0267). *(AIM Para 222)*
Given a wind as shown by the landing direction indicator in Fig. 28, the pilot should land to the south on Runway 18 and expect a crosswind from the right.

Answer (A) is incorrect because the wind is from the southwest. The landing should be into the wind. Answers (C) and (D) are incorrect because Runways 4 and 22 are closed.

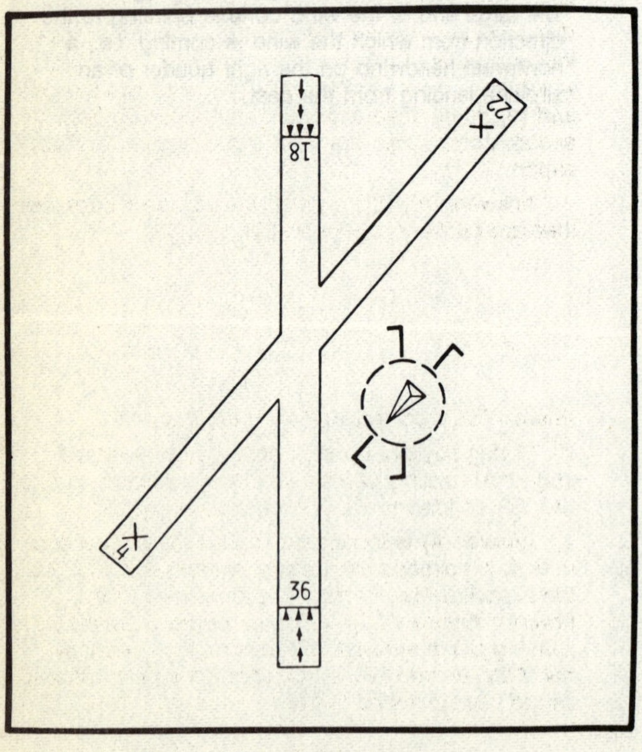

FIGURE 28.—Airport Diagram.

Chapter 5: Airports and Air Traffic Control

12.
0268. (Refer to figure 28.) The arrows that appear on the end of the north/south runway indicate that the area

A— may be used only for taxiing.
B— is usable for taxiing, takeoff, and landing.
C— cannot be used for landing, but may be used for taxiing and takeoff.
D— is available for landing at the pilot's discretion.

Answer (C) is correct (0268). *(AIM Para 60)*
The arrows that appear on the end of the north/south runway (displaced thresholds), as shown in Fig. 28, indicate that the area cannot be used for landing, but may be used for taxiing and takeoff.
Answer (A) is incorrect because takeoffs as well as taxiing are permitted. Answers (B) and (D) are incorrect because landings are not permitted on displaced thresholds.

13.
0266. (Refer to figure 28.) Select the proper traffic pattern and runway for a landing as indicated on the airport diagram.

A— Right-hand traffic and Rwy 4.
B— Right-hand traffic and Rwy 18.
C— Left-hand traffic and Rwy 22.
D— Left-hand traffic and Rwy 36.

Answer (B) is correct (0266). *(AIM Para 222)*
The tetrahedron indicates wind direction by pointing into the wind. On Fig. 28, Runways 4 and 22 are closed, as indicated by the "X." Accordingly, with the wind from the southwest, the landing should be made on Runway 18. Runway 18 has right-hand traffic, as indicated by the leg extending out of the segmented circle on its south side.
Answer (A) is incorrect because Runway 4 is closed as indicated by the "X." Answer (C) is incorrect because the "X" markings indicate that Runways 4 and 22 are closed. Answer (D) is incorrect because the landing on Runway 36 would be downwind and would result in a left-quartering tailwind.

14.
0367. Which is the correct traffic pattern departure procedure to use at a noncontrolled airport?

A— Depart in any direction consistent with safety, after crossing the airport boundary.
B— Make all turns to the left.
C— Comply with any FAA traffic pattern established for the airport.
D— Depart as prearranged with other pilots using the airport.

Answer (C) is correct (0367). *(FAR 91.89)*
Each person operating an airplane to or from an airport without an operating control tower shall (a) in the case of an airplane approaching to land, make all turns of that airplane to the left unless the airport displays approved light signals or visual markings indicating that turns should be made to the right, in which case the pilot shall make all turns to the right, and (b) in the case of an airplane departing the airport, comply with any FAA traffic pattern for that airport.
Answers (A), (B), and (D) are incorrect because they are contrary to FAR 91.89.

BEACONS

15.
0254. An airport's rotating beacon operated during the daylight hours indicates

A— there are obstructions on the airport.
B— that weather in the control zone is below basic VFR weather minimums.
C— parachute jumping is in progress.
D— the airport is temporarily closed.

Answer (B) is correct (0254). *(AIM Para 48)*
During daylight hours, operation of the green and white rotating beacon indicates that the airport has IFR conditions if it is a controlled airport.
Answer (A) is incorrect because the obstructions near or on airports are usually listed in NOTAMS or the airport/facility directory appropriate to their hazard. Answer (C) is incorrect because parachute jumping is not indicated by beacon, but rather by NOTAM. Answer (D) is incorrect because a closed airport has no airport lighting.

UNICOM AND MULTICOM

16.
0279. When contacting airport UNICOM facilities, which CTAF MHz frequencies are commonly used?

A— 122.6, 122.8, 123.0.
B— 122.7, 122.8, 123.0.
C— 125.5, 122.8, 123.0.
D— 125.6, 122.8, 123.0.

Answer (B) is correct (0279). *(AIM Para 159)*
UNICOM frequencies, which act as Common Traffic Advisory Frequencies (CTAF) at uncontrolled airports, are 122.7, 122.8, and 123.0.
Answer (A) is incorrect because 122.6 is a Flight Service Station frequency. Answers (C) and (D) are incorrect because 125.5 and 125.6 are ATC frequencies.

17.
0280. In the event an airport facility has no tower, FSS, or UNICOM, pilots may self-announce their intentions on MULTICOM frequency

A— 122.9.
B— 123.2.
C— 123.4.
D— 128.3.

Answer (A) is correct (0280). *(AIM 157, 159)*
At airports where there is no tower, no Flight Service Station, or no UNICOM, use the MULTICOM frequency 122.9 for announcing your position and intentions. When inbound, you should call approximately 10 miles from the airport and state your aircraft identification and type, altitude, location relative to the airport, and intentions. Report again on the downwind, base, and final approach legs as appropriate.
Answers (B) and (C) are incorrect because 123.2 and 123.4 are reserved for Flight Service Stations. Answer (D) is incorrect because 128.3 is reserved for ATC.

VISUAL APPROACH SLOPE INDICATORS (VASIs)

18.
0259. (Refer to figure 26.) Illustration A indicates that an aircraft is

A— off course.
B— below the glide slope.
C— on the glide slope.
D— above the glide slope.

Answer (C) is correct (0259). *(AIM Para 41)*
Illustration "A" indicates that the airplane is on the glidepath (glide slope). The red row of lights is above the white row.
Answer (A) is incorrect because, if the airplane is excessively off course, the VASI system would not be visible. Additionally, VASI has nothing to do with runway alignment. Answer (B) is incorrect because, if the airplane is below the glidepath, both rows of lights will be red, as indicated in "D."
Answer (D) is incorrect because, if the aircraft is above the glidepath, both lights will be white, as indicated in "C."

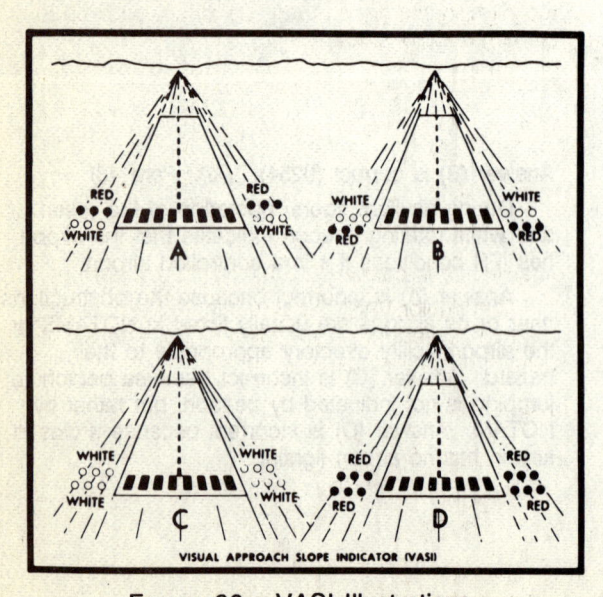

FIGURE 26.—VASI Illustrations.

Chapter 5: Airports and Air Traffic Control

19.
0260. (Refer to figure 26.) While on final approach to a runway equipped with a standard 2-bar VASI, the lights appear as shown in illustration D. This means that the pilot is

A— receiving an erroneous light indication.
B— above the glide slope.
C— below the glide slope.
D— on the glide slope.

20.
0261. (Refer to figure 26.) VASI lights appearing as shown in illustration C would indicate that an airplane is

A— off course to the left.
B— on the glide slope.
C— below the glide slope.
D— above the glide slope.

21.
0256. An on glide slope indication from a tri-color VASI is

A— a white light signal.
B— a green light signal.
C— an amber light signal.
D— a pink light signal.

22.
0257. An above glide slope indication from a tri-color VASI is

A— a white light signal.
B— a green light signal.
C— an amber light signal.
D— a pink light signal.

23.
0258. A below glide slope indication from a tri-color VASI is

A— a pink light signal.
B— a green light signal.
C— an amber light signal.
D— a red light signal.

Answer (C) is correct (0260). *(AIM Para 41)*
In Illustration "D", both rows of lights are red. Thus, the aircraft is below the glidepath. Remember, "red means dead."
Answer (A) is incorrect because Illustration "B" is an example of an erroneous (i.e., impossible) light indication. White cannot appear over red. Answer (B) is incorrect because, if the airplane is above the glidepath, the lights would both show white, as indicated in "C." Answer (D) is incorrect because, if the airplane is on the glidepath, the lights would be red over white, as indicated in "A."

Answer (D) is correct (0261). *(AIM Para 41)*
In Illustration "C," all light bars are white, which means the airplane is above the glidepath.
Answer (A) is incorrect because the VASI does not alert a pilot as to runway alignment, but a pilot who is excessively to the left or right would be unable to see the VASI lights at all. Answer (B) is incorrect because, if the pilot is on the glidepath, the lights would appear red over white, as indicated in "A." Answer (C) is incorrect because, if the airplane is below the glidepath, both lights would show red, as indicated in "D."

Answer (B) is correct (0256). *(AIM Para 42)*
Tricolor visual approach slope indicators normally consist of a single light unit projecting a three-color visual approach path into the final approach area of the runway, upon which the indicator is installed. The below glidepath indicator is red. The above glidepath indicator is amber. The on glidepath indicator is green. This type of indicator has a useful range of approximately ½ to 1 mile in daytime and up to 5 miles at night.

Answer (C) is correct (0257). *(AIM Para 42)*
The tricolor VASI has three lights: amber for too high, green for on the glide slope, and red for below the glide slope.

Answer (D) is correct (0258). *(AIM Para 42)*
The tricolor VASI has three lights: amber for too high, green for on the glide slope, and red for below the glide slope.

AIRPORT TRAFFIC AREAS

24.
0364. What is the purpose of an Airport Traffic Area?

A— To provide for the control of aircraft landing and taking off from an airport with an operating control tower.
B— To provide for the control of all aircraft operating in the vicinity of an airport with an operating control tower.
C— To provide for the control of air traffic within a control zone that has an operating control tower.
D— To restrict aircraft without radios from operating in the vicinity of an airport with an operating control tower.

Answer (A) is correct (0364). *(FAR 91.85, 91.87)*
Airport traffic areas control airplanes landing and taking off from an airport with an operating control tower.
Answer (B) is incorrect because FAR 91.87 provides that airport traffic areas can only control those airplanes operating to, from, or on airports with an operating control tower. Answer (C) is incorrect because control zones relate to more stringent VFR weather minimums and IFR operations. Answer (D) is incorrect because an airplane without two-way communications may operate within an airport traffic area, but may not land or take off.

25.
0365. When is an Airport Traffic Area in effect?

A— From 1 hour before sunrise to 1 hour after sunset.
B— When the associated control tower is in operation.
C— When the associated FSS is in operation.
D— From sunrise to sunset.

Answer (B) is correct (0365). *(FAR 1.1)*
An airport traffic area is automatically in effect when and only when its associated control tower is in operation regardless of weather conditions, availability of radar services, or time of day. Airports with "part-time" operating towers only have "part-time" airport traffic areas.

26.
0313. An Airport Traffic Area is automatically in effect when

A— its associated control tower is in operation.
B— the weather is below VFR minimums.
C— nighttime hours exist.
D— radar service is available.

Answer (A) is correct (0313). *(FAR 1.1)*
An airport traffic area is in effect if the control tower is open irrespective of the weather, daylight conditions, or available radar.

27.
0276. What are the horizontal limits of an Airport Traffic Area?

A— 3 SM from the airport boundary.
B— 3 SM from the geographical center of the airport.
C— 5 SM from the airport boundary.
D— 5 SM from the geographical center of the airport.

Answer (D) is correct (0276). *(FAR 1.1)*
An airport traffic area is the airspace within a horizontal radius of 5 statute miles from the geographical center of any airport at which a control tower is operating.

Chapter 5: Airports and Air Traffic Control

28.
0277. The vertical limit of an Airport Traffic Area is from the surface up to

A— and including 1,500 feet.
B— and including 2,000 feet.
C— but not including 3,000 feet.
D— the base of the overlying control area.

Answer (C) is correct (0277). *(FAR 1.1)*
An airport traffic area extends from the surface up to, but not including, 3,000 ft above the surface.

AIRPORT ADVISORY AREAS

29.
0271. An FSS located at an airport without a control tower provides advisory service

A— only for aircraft remaining in the traffic pattern.
B— only for departing aircraft.
C— only for arriving aircraft.
D— for both departing and arriving aircraft.

Answer (D) is correct (0271). *(AIM Para 131)*
Airport Advisory Areas encompass an area within a 10 statute mile radius of an airport where a control tower is not operating but a Flight Service Station is located. The FSS provides advisory service to arriving and departing aircraft on a voluntary rather than a mandatory basis.

Answers (A), (B) and (C) are incorrect because the advisory service is for all participating aircraft within a 10-mile radius of the airport.

30.
0278. Prior to entering an Airport Advisory Area, a pilot

A— must obtain a clearance from Air Traffic Control.
B— should monitor ATIS for weather and traffic advisories.
C— should contact approach control for vectors to the traffic pattern.
D— should contact the local FSS for airport and traffic advisories.

Answer (D) is correct (0278). *(AIM Para 157)*
Airport Advisory Areas exist at noncontrolled airports that have a Flight Service Station (FSS). The FSS provides advisory (not control) information on traffic, weather, etc. to requesting aircraft. Accordingly, pilots should (not must) contact FSSs for advisory services.

Answer (A) is incorrect because airport advisory areas are nonmandatory and thus require no special ATC clearances. Answer (B) is incorrect because automatic terminal information service (ATIS) provides prerecorded weather and airport data but not traffic advisories. Answer (C) is incorrect because FSS advisory areas (immediately around airports) do not exist where there is approach control.

TCAs AND AIRPORT RADAR SERVICE AREAS

31.
0314. The normal radius of the outer area of an Airport Radar Service Area is

A— 5 SM.
B— 5 NM.
C— 15 NM.
D— 20 NM.

Answer (D) is correct (0314). *(AIM Para 100)*
The outer area (not outer circle) of an ARSA has a normal radius of 20 NM with variations possible. This is in contrast to the ARSA itself, which also has two circles: the first of 5 NM and the second of 10 NM. Note that the question asked for the outer area radius of an ARSA.

Answer (A) is incorrect because ARSA dimensions are in NM, not SM. Answer (B) is incorrect because 5 NM is the radius of the inner circle of the ARSA itself. Answer (C) is incorrect because 15 NM is not an ARSA dimension.

85

Chapter 5: Airports and Air Traffic Control

32.
0315. The vertical limit of an Airport Radar Service Area is

A— 1,200 feet AGL.
B— 3,000 feet AGL.
C— 4,000 feet above the primary airport.
D— up to, but not including 14,500 feet.

Answer (C) is correct (0315). *(AIM Para 100)*
The vertical limit (ceiling) of both bands in ARSAs is 4,000 ft AGL.
Answer (A) is incorrect because 1,200 ft AGL is the base of the outer circle of the ARSA (5-10 NM from primary airport). Answer (B) is incorrect because 3,000 ft AGL is the ceiling of airport traffic areas. Answer (D) is incorrect because 14,500 ft is the base of the continental control area.

33.
0368. When flying a transponder-equipped aircraft and unless otherwise advised, a recreational pilot should squawk which VFR code number?

A— 1200.
B— 1250.
C— 1500.
D— 1520.

Answer (A) is correct (0368). *(AIM Para 170)*
Unless otherwise instructed by ATC, your transponder should be set 1200, regardless of altitude, when flying VFR.
Answers (B), (C) and (D) are incorrect because none have any special significance and may be assigned by ATC. Code 7500 is "Hijack." Code 7600 is "Lost Radio Communications." Code 7700 is to indicate an emergency. Code 4000 is for military pilots within restricted/warning areas. Code 7777 is "Military Interceptor Operations."

34.
0283. ARSA and TCA controlled airspace geographical limits are depicted

A— only on Sectional Aeronautical Charts.
B— only on VFR Terminal Area Charts.
C— on VFR Terminal Area Charts and Sectional Aeronautical Charts.
D— on NOAA Radar Charts, VFR Terminal Area Charts, and Sectional Aeronautical Charts.

Answer (C) is correct (0283). *(AIM Paras 97, 100)*
ARSAs are charted on sectional charts and VFR terminal area charts. TCAs are charted on sectional, world aeronautical, and terminal area charts.
Answer (A) is incorrect because they are also shown on other charts, such as the VFR terminal area charts. Answer (B) is incorrect because they are also shown on other charts such as sectional charts. Answer (D) is incorrect because NOAA radar charts are radar summary (weather) charts.

CHAPTER SIX
WEIGHT AND BALANCE

This chapter contains outlines of major concepts tested, all 16 FAA test questions regarding weight and balance, and an explanation of each answer.

CAUTION: Recall that the <u>sole purpose</u> of this book is to expedite your passing the FAA written test for the recreational pilot certificate. Accordingly, all extraneous material (i.e., topics or regulations not directly tested on the FAA written test) is omitted, even though much more information and knowledge are necessary to fly safely. This additional material is presented in *RECREATIONAL PILOT FLIGHT MANEUVERS* and *PRIVATE PILOT HANDBOOK*, available from Aviation Publications, Inc. See pages 208 to 210 for more information and an order form.

Most of the questions in this chapter require interpretation of graphs and charts. Graphs and charts pictorially describe the relationship between two or more variables. Thus, they are a substitute for solving one or more equations. Each time you must interpret (i.e., get an answer from) a graph or chart, you should

1. Understand clearly what is required, e.g., landing roll distance, range in hours, load factor, etc.
2. Analyze the chart or graph to determine the variables involved, including
    a. Labeled sides (axes) of the graph or chart.
    b. Labeled lines within the graph or chart.
3. Plug the data given in the question into the graph or chart.
4. Finally, determine the value of the item required in the question.

> *A thorough, illustrated explanation of weight and balance appears on pages 127-140 of Private Pilot Handbook (Third Edition).*

WEIGHT AND BALANCE (16 questions)

1. Airplanes must be loaded so that the center of gravity will be forward of the center of lift. This provides airplane stability about the lateral axis (i.e., pitch) and ensures that the nose will drop if the wing stalls.

2. Unusable fuel and optional equipment are included in the *basic empty weight* of an airplane. Basic empty weight also includes hydraulic fluid and undrainable oil (and, in some airplanes, all of the oil).
    a. When working weight and balance questions on the exam, note if oil has been specified as included in the empty weight.

3. When working with fuel weights (1 gallon = 6 lb).

4. The *center of gravity (CG)* is a point of fore and aft balance in an airplane. Its location is determined by multiplying the weight of the airplane, and everything that is put into the airplane, by its distance (either positive or negative) from an imaginary plane, or line, called a *reference datum*. The datum may be inside or outside of the airplane. The distance between each item and the datum is called the *arm*. The CG location can be calculated mathematically (see question 4 on page 90) but is usually found by using graphs (see Figure 16 on page 93) or charts (see pages 96-97 and questions 10-16).

5. The *loading graph* (see Figure 16 on page 93) is used to determine the total aircraft moment. The moment is the product of the total weight of the airplane with all of its contents, and the distance from the datum to the point in the airplane where all of that weight is centered.

   a. On most loading graphs, weights in pounds are listed along the left side. Using Figure 16 on page 93, select each item's weight from the scale along the left side of the graph, then move horizontally to intersect the diagonal line appropriate to that item. (Diagonal lines are usually provided for pilot, passengers, fuel, oil, baggage, etc.)

   b. From the point of intersection of the horizontal weight line and the appropriate diagonal line, move straight down to the scale along the bottom of the graph where the moments are shown. The moment scale is divided by 1,000 (moment ÷ 1,000) to make the numbers smaller and thereby more manageable.

      1) Note that you may have to estimate some moments when it is not clear exactly where the diagonal line intersects, e.g., the pilot and front seat passenger diagonal at 400 pounds intersects somewhere between 14.5 and 15.0 moments. Do not let this worry you, as using 15.0 will be close enough.

   c. Total the weights and moments.

   d. EXAMPLE. Determine the center of gravity moment/1,000 pounds-inches given the following situation. First, set up a schedule of what you are given and what you must find:

	Weight (lb)	Moment/1000 lb inches
Empty weight	1,364	51.7
Pilot and front seat passenger	400	?
Baggage	120	?
Fuel (38 gal usable)	228	11.0
Oil (8 qt)	15	-0.2

      1) Compute the moment of the pilot and front seat passenger by referring to the loading graph and locate 400 pounds on the weight scale. Move horizontally across the graph to intersect the diagonal line representing the pilot and front passenger, and then to the bottom scale which indicates a moment of approximately 15.0.

      2) Locate 120 on the weight scale for the baggage. Move horizontally across the graph to intersect the diagonal line that represents baggage, then down vertically to the bottom which indicates a moment of approximately 11.5.

      3) Notice a negative 0.2 moment for the engine oil. Add all moments except this negative moment and obtain a total of 89.2. Then subtract the negative moment to obtain a total moment of 89.0.

      4) You should also add all the weights to determine that the maximum gross weight is not exceeded.

	Weight (lb)	Moment/1000 lb inches
Empty weight	1,364	51.7
Pilot and front seat passenger	400	15.0
Baggage	120	11.5
Fuel (38 gal usable)	228	11.0
Oil (8 qt)	15	-0.2
	2,127	89.0

Chapter 6: Weight and Balance

6. The *center of gravity moment envelope chart* is a graph showing CG limits for various gross weights, i.e., acceptable limits are established as an area with weight on the vertical axis and moments on the horizontal axis. See Figure 16 on page 93. To determine if the total airplane weight and the moment/1,000 "pounds-inches" that you have calculated is within allowable limits:

   a. Identify the point on the center of gravity moment envelope graph by first plotting the total loaded aircraft weight across to the right from the column of weights along the left edge.

   b. Plot the moment upward from the bottom.

   c. The intersection where the horizontal weight line and the vertical moment line cross lies either within or outside the center of gravity (CG) moment envelope (the boxes drawn on the graph).

   d. EXAMPLE. Using the data in the example above, locate the weight of 2,127 pounds on the vertical axis and then move across the chart to the moment line of 89.0. The point of intersection indicates that the airplane is within both CG and gross weight limits.

   e. When the loaded airplane moment falls within the utility category envelope, the FAA permits certain maneuvers that are precluded in the normal category.

7. The location of the CG itself in inches from the datum is Total Moment/Total Weight.

   a. For the example above, CG = 89,000 ÷ 2,127 = 41.8 inches aft of the datum.

   b. Be careful to check whether the required answer is in moments or inches or pounds.

8. Without graphs, you need a calculator to compute the moments. Since this approach is not tested on the written exam, it is not illustrated here.

QUESTIONS AND ANSWER EXPLANATIONS

All of the FAA questions from the written test for the Recreational Pilot certificate relating to weight and balance and the material outlined previously are reproduced below in the same modules as the previous outlines. To the immediate right of each question are the correct answer and answer explanation. You should cover these answers and answer explanations while responding to the questions. Refer to the general discussion in Chapter 1 on how to take the examination.

Remember that the questions from the FAA's Written Test Book have been reordered by topic, and the topics have been organized into a meaningful sequence. Accordingly, the first line of the answer explanation gives the FAA question number and the citation of the authoritative source for the answer.

Note that most of these questions require much more time to work than the few that are more definitional in nature. Remember on the exam that you have 4 hours to work 50 questions, which is an average of about 5 minutes each. Since most questions can be answered in 2 minutes, you may thus spend as much as 10 minutes each on a few questions and still have plenty of time.

1.
0158. An aircraft is loaded 110 pounds over maximum certificated gross weight. If fuel (gasoline) is drained to bring the aircraft weight within limits, how much fuel should be drained?

A— 15.7 gallons.
B— 16.2 gallons.
C— 17.1 gallons.
D— 18.4 gallons.

Answer (D) is correct (0158). *(PHAK Chap IV)*
Fuel weighs 6 lb/gal. If an airplane is 110 lb over maximum gross weight, 18.4 gals (110 lb ÷ 6) must be drained to bring the airplane weight within limits. Note that the effect on total moment is not required.

2.
0157. Which items are included in the basic empty weight of an airplane?

A— Hydraulic fluid and usable fuel.
B— Only the airframe, powerplant, and equipment installed by the manufacturer.
C— Full fuel tanks and engine oil to capacity, but excluding crew and baggage.
D— Unusable fuel and optional equipment.

Answer (D) is correct (0157). *(PHAK Chap IV)*
The basic empty weight of an airplane includes airframe, engines, and all items of operating equipment that have fixed locations and are permanently installed. It includes optional and special equipment, fixed ballast, hydraulic fluid, unusable fuel, and undrainable oil.

Answers (A) and (C) are incorrect because usable fuel (included in full fuel) is not a component of basic empty weight. Answer (B) is incorrect because unusable and undrainable fuel and oil and optional equipment permanently installed are also included.

3.
0171. If an aircraft is loaded 90 pounds over maximum certificated gross weight and fuel (gasoline) is drained to bring the aircraft weight within limits, how much fuel should be drained?

A— 6 gallons.
B— 9 gallons.
C— 12 gallons.
D— 15 gallons.

Answer (D) is correct (0171). *(PHAK Chap IV)*
Since fuel weighs 6 lb/gal, draining 15 gal (90 lb ÷ 6) will reduce the weight of an airplane that is 90 lb over maximum gross weight to the acceptable amount.

4.
0172. GIVEN:

	WEIGHT (LB)	ARM (IN)	MOMENT (LB/IN)
Empty weight	1,495.0	101.4	151,593.0
Pilot and passenger	380.0	64.0	...
Fuel (30 gal usable no reserve)	...	96.0	...

The CG is located how far aft of datum?

A— 92.44.
B— 94.01.
C— 119.8.
D— 135.0.

Answer (B) is correct (0172). *(PHAK Chap IV)*
To compute the CG you must first multiply each weight by the arm to get the moment. Note that the fuel is given as 30 gal. To get the weight, multiply the 30 by 6 lb/gal (30 × 6) = 180 lb.

	Weight (lb)	Arm (in)	Moment (lb in)
Empty weight	1,495.0	101.4	151,593.0
Pilot and passengers	380.0	64.0	24,320.0
Fuel (30 × 6)	180.0	96.0	17,280.0
	2,055.0		193,193.0

Now add the weights and moments. To get CG, you divide total moment by total weight (193,193 ÷ 2,055.0) = a CG of 94.01 inches.

Chapter 6: Weight and Balance

Figure 16 is presented on page 93.

5.
0174. (Refer to figure 16.) What is the maximum load that may be carried in the baggage area for the weight and CG to remain within the loading envelope?

	WEIGHT (LB)	MOM/1000
Empty weight	1,350	51.5
Pilot and front passenger	250	...
Baggage	...	...
Fuel, 48 gal	...	...
Oil, 8 qt	15	-0.2

A— 90 pounds.
B— 105 pounds.
C— 120 pounds.
D— 277 pounds.

Answer (C) is correct (0174). *(PHAK Chap IV)*

To compute the amount of weight left for baggage, compute each individual moment by using the loading graph and add them up. First, compute the moment for the pilot and front seat passenger with a weight of 250 lb. Refer to the loading graph and the vertical scale at the left side to find the value of 250. From this position, move to the right across the graph until you intersect the diagonal line that represents pilot and front passenger. From this point, move vertically down to the bottom scale, which indicates a moment of about 9.2.

To compute the moment of the fuel, you must recall that fuel weighs 6 lb/gal. The question gives 48 gal for a total fuel weight of 288 lb. Now move up the weight scale on the loading graph to 288, then horizontally across to intersect the diagonal line that represents fuel, then vertically down to the moment scale, which indicates approximately 13.6.

Now total the weights before baggage (1,903 lb including 15 lb of engine oil). Also total the moments (74.1 including engine oil with a negative 0.2 moment).

With this information, refer to the center of gravity moment envelope chart. Note that the maximum weight in the envelope is 2,300 lb, but the maximum in the baggage area is 120 lb. Recompute the total weight and moments with 120 lb to determine that the total weight of 2,023 will fall within the CG envelope.

The total of 85.6 moments is within the CG envelope at 2,023 lb of weight. Therefore, baggage of 120 lb can be loaded.

	Weight	Moment/1000 lb inches
Empty weight	1,350	51.5
Pilot and front seat passenger	250	9.2
Baggage	120	11.5
Fuel (48 gal x 6 lb/gal)	288	13.6
Oil (8 qt)	15	-0.2
	2,023	85.6
		(with 120 lb baggage)

Chapter 6: Weight and Balance

6.
0175. (Refer to figure 16.) Calculate the loaded aircraft moment/1000 of the airplane loaded as follows and determine which category when plotted on the CG moment envelope.

	WEIGHT (LB)	MOM/1000
Empty weight	1,350	51.5
Pilot and front passenger	310	...
Fuel, 38 gal	...	...
Oil, 8 qt	15	-0.2

A— 73.8, utility category.
B— 79.2, utility category.
C— 80.8, normal category.
D— 81.2, normal category.

Answer (A) is correct (0175). *(PHAK Chap IV)*
First, total the weight and get 1,903 lb. Note that the 38 gal of fuel weighs 228 lb (38 gal x 6 lb/gal).

Find the moments for the pilot and front seat passengers and fuel by using the loading graph as explained in the answer to question 5. Total the moments as shown in the schedule below.

Now refer to the center of gravity moment envelope. Find the gross weight of 1,903 on the vertical scale, and move horizontally across the chart until intersecting the vertical line that represents the 73.8 moment. Note that a moment of 73.8 pound-inches falls into the utility category envelope.

	Weight	Moment/1000 lb inches
Empty weight	1,350	51.5
Pilot and front seat passenger	310	11.5
Fuel (38 gal x 6 lb/gal)	228	11.0
Oil (8 qt)	15	-0.2
	1,903	73.8

7.
0176. (Refer to figure 16.) What is the maximum amount of fuel that may be aboard the airplane on takeoff if it is loaded as follows?

	WEIGHT (LB)	MOM/1000
Empty weight	1,350	51.5
Pilot and front passenger	340	...
Baggage	120	...
Oil, 8 qt	15	...

A— 24 gallons.
B— 34 gallons.
C— 48 gallons.
D— 79 gallons.

Answer (C) is correct (0176). *(PHAK Chap IV)*
To find the maximum amount of fuel this airplane can carry, add the empty weight (1,350), pilot and front passenger weight (340), baggage (120), and oil (15), for a total of 1,825 lb. Gross weight maximum on the center of gravity moment envelope chart is 2,300. Thus, 475 lb of weight (2,300 - 1,825) is available for fuel. The maximum fuel is 48 gal (288 lb) if its center of gravity moments do not exceed the limit. Note that long-range tanks were not mentioned; assume they exist.

Compute the moments for each item as explained in the answer to question 5. The empty weight moment is given as 51.5. Calculate the moment for the pilot and front passenger as 12.5, the fuel as 13.6, the baggage as 11.5, and the oil as -0.2. These total to 88.9, which is within the envelope, so 48 gal of fuel may be carried.

	Weight	Moment/1000 lb inches
Empty weight	1,350	51.5
Pilot and front seat passenger	340	12.5
Baggage	120	11.5
Fuel (48 gal x 6 lb/gal)	288	13.6
Oil (8 qt)	15	-0.2
	2,113	88.9

Chapter 6: Weight and Balance

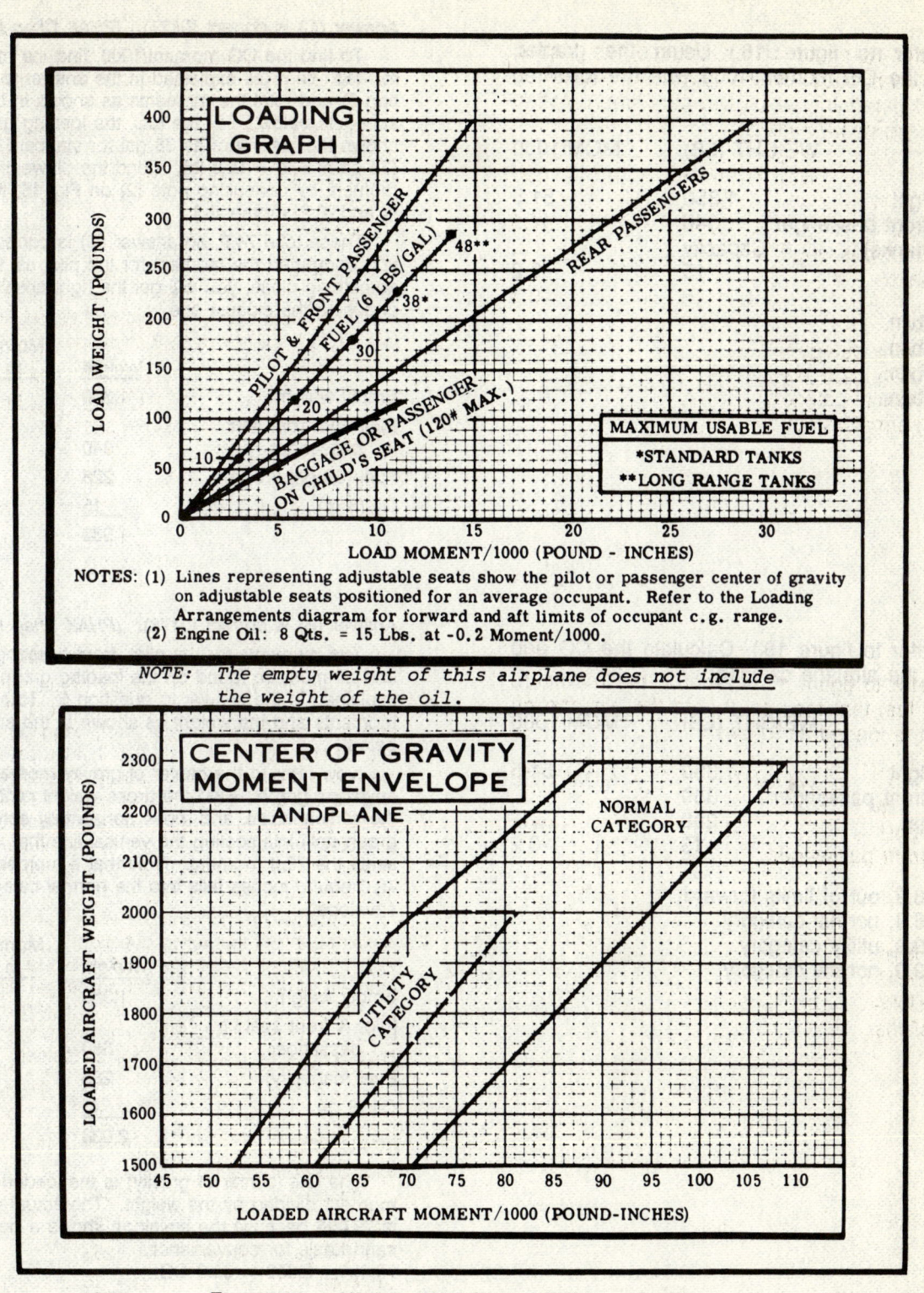

FIGURE 16.—Airplane Weight and Balance Graphs.

94 Chapter 6: Weight and Balance

Figure 16 is presented on page 93.

8.
0177. (Refer to figure 16.) Using the graphs, determine the CG moment/1000 with the following data.

	WEIGHT (LB)	MOM/1000
Empty weight	1,350	51.5
Pilot and front passenger	340	...
Fuel (std. tanks)	Capacity	...
Oil, 8 qt	15	-0.2

A— 38.7 lb/in.
B— 69.9 lb/in.
C— 74.9 lb/in.
D— 77.5 lb/in.

Answer (C) is correct (0177). *(PHAK Chap IV)*

To find the CG moment/1000, find the moments for each item (as explained in the answer to question 5) and total the moments as shown in the schedule below. For the fuel, the loading graph shows the maximum as 38 gal for standard tanks (38 gal x 6 lb = 228 lb). (Find the oil weight and moment by consulting note (2) on Fig. 16; it is 15 lb and -0.2. moments.)

These total 74.9, so answer (C) is correct. If you computed the moment for the pilot as 12.6 or something close, you will get the right answer by choosing the closest one.

	Weight	Moment/1000 lb inches
Empty weight	1,350	51.5
Pilot and front seat passenger	340	12.6
Fuel (38 gal x 6 lb)	228	11.0
Oil (8 qt)	15	-0.2
	1,933	74.9

9.
0178. (Refer to figure 16.) Calculate the CG and determine the airplane category.

	WEIGHT (LB)	MOM/1000
Empty weight	1,350	51.5
Pilot and front passenger	380	...
Fuel, 48 gal	288	...
Oil, 8 qt	15	-0.2

A— CG 38.9, out of limits forward.
B— CG 38.9, normal category.
C— CG 39.9, utility category.
D— CG 39.9, normal category.

Answer (B) is correct (0178). *(PHAK Chap IV)*

The moments for the pilot, front passenger, fuel, and oil must be found on the loading graph as explained in the answer to question 5. Total all the moments and the weight as shown in the schedule below.

Now refer to the center of gravity moment envelope graph. Find the gross weight of 2,033 on the vertical scale, and move horizontally across the graph until intersecting the vertical line that represents the 79.0 moment. Note that a moment of 79.0 pound-inches falls into the normal category envelope.

	Weight	Moment/1000 lb inches
Empty weight	1,350	51.5
Pilot & front seat passenger	380	14.0
Fuel (capacity)	288	13.7
Oil (8 qt)	15	-0.2
	2,033	79.0

The CG (center of gravity) is the loaded airplane moment divided by the weight. The actual moment is 79,000 because the envelope shows a per thousand basis for convenience.
79,000 ÷ 2,033 = 38.9 CG.

Chapter 6: Weight and Balance

Figures 19 and 20 are presented on pages 96 and 97.

10.
0187. (Refer to figures 19 and 20.) Determine if the airplane weight and balance are within limits.

Front seat occupants .... 340 lb
Rear seat cargo .... 295 lb
Fuel .... 44 gal
Baggage .... 56 lb

A— Within limits.
B— Weight within limits, CG out of limits forward.
C— 20 pounds overweight, CG within limits.
D— 39 pounds overweight, CG out of limits forward.

Answer (C) is correct (0187). *(PHAK Chap IV)*

Both the total weight and the total moment must be calculated. As in most weight and balance problems, you should begin by setting up a schedule as below. Note that the empty weight in Fig. 19 is given as 2,015 with a moment per 100 inches of 1,554 (note the change to "per 100 inches" on this chart), and that empty weight includes the oil.

The next step is to compute the moment for each of the CG positions. The front seat moment of 340 lb is twice that for 170 lb, i.e., 2 x 144 = 288.

The rear seat moment for 295 lb can be computed by adding the moments for 150 and 145 lb. The moment for 150 lb is 182. Interpolate the moment for 145 lb by splitting the difference between 169 and 182, i.e., add 6 to the 169 moment of 140 to get 175. Thus, the total rear seat moment is 357 (182 + 175).

For the 264 lb of fuel, the moment is 198. For 56 lb of baggage, the moment is approximately 78 (interpolation required again). Add up the moments to 2,475.

The last step is to go to the Moment Limits versus Weight chart (Fig. 20), and note that the maximum weight allowed is 2,950, which means that the plane is 20 lb over. At 2,475, the plane is within the CG limits because the moments may be from 2,422 to 2,499 at 2,950 lb.

	Weight	Moment/100 lb inches
Empty weight with oil	2,015	1,554
Front seat	340	288
Rear seat	295	357
Fuel (44 gal x 6 lb/gal)	264	198
Baggage	56	78
	2,970	2,475

USEFUL LOAD WEIGHTS AND MOMENTS

OCCUPANTS

FRONT SEATS ARM 85		REAR SEATS ARM 121	
Weight	Moment/100	Weight	Moment/100
120	102	120	145
130	110	130	157
140	119	140	169
150	128	150	182
160	136	160	194
170	144	170	206
180	153	180	218
190	162	190	230
200	170	200	242

BAGGAGE OR 5TH SEAT OCCUPANT ARM 140

Weight	Moment/100
10	14
20	28
30	42
40	56
50	70
60	84
70	98
80	112
90	126
100	140
110	154
120	168
130	182
140	196
150	210
160	224
170	238
180	252
190	266
200	280
210	294
220	308
230	322
240	336
250	350
260	364
270	378

USABLE FUEL

MAIN WING TANKS ARM 75

Gallons	Weight	Moment/100
5	30	22
10	60	45
15	90	68
20	120	90
25	150	112
30	180	135
35	210	158
40	240	180
44	264	198

AUXILIARY WING TANKS ARM 94

Gallons	Weight	Moment/100
5	30	28
10	60	56
15	90	85
19	114	107

*OIL

Quarts	Weight	Moment/100
10	19	5

*Included in Basic Empty Weight

Empty Weight – 2015

MOM/100 – 1554

MOMENT LIMITS vs WEIGHT

Moment limits are based on the following weight and center of gravity limit data (landing gear down).

WEIGHT CONDITION	FORWARD CG LIMIT	AFT CG LIMIT
2950 lb. (take-off or landing)	82.1	84.7
2525 lb.	77.5	85.7
2475 lb. or less	77.0	85.7

FIGURE 19.—Airplane Weight and Balance Tables.

Chapter 6: Weight and Balance

MOMENT LIMITS vs WEIGHT (Continued)

Weight	Minimum Moment 100	Maximum Moment 100	Weight	Minimum Moment 100	Maximum Moment 100
2100	1617	1800	2600	2037	2224
2110	1625	1808	2610	2048	2232
2120	1632	1817	2620	2058	2239
2130	1640	1825	2630	2069	2247
2140	1648	1834	2640	2080	2255
2150	1656	1843	2650	2090	2263
2160	1663	1851	2660	2101	2271
2170	1671	1860	2670	2112	2279
2180	1679	1868	2680	2123	2287
2190	1686	1877	2690	2133	2295
2200	1694	1885	2700	2144	2303
2210	1702	1894	2710	2155	2311
2220	1709	1903	2720	2166	2319
2230	1717	1911	2730	2177	2326
2240	1725	1920	2740	2188	2334
2250	1733	1928	2750	2199	2342
2260	1740	1937	2760	2210	2350
2270	1748	1945	2770	2221	2358
2280	1756	1954	2780	2232	2366
2290	1763	1963	2790	2243	2374
2300	1771	1971			
2310	1779	1980	2800	2254	2381
2320	1786	1988	2810	2265	2389
2330	1794	1997	2820	2276	2397
2340	1802	2005	2830	2287	2405
2350	1810	2014	2840	2298	2413
2360	1817	2023	2850	2309	2421
2370	1825	2031	2860	2320	2428
2380	1833	2040	2870	2332	2436
2390	1840	2048	2880	2343	2444
2400	1848	2057	2890	2354	2452
2410	1856	2065	2900	2365	2460
2420	1863	2074	2910	2377	2468
2430	1871	2083	2920	2388	2475
2440	1879	2091	2930	2399	2483
2450	1887	2100	2940	2411	2491
2460	1894	2108	2950	2422	2499
2470	1902	2117			
2480	1911	2125			
2490	1921	2134			
2500	1932	2143			
2510	1942	2151			
2520	1953	2160			
2530	1963	2168			
2540	1974	2176			
2550	1984	2184			
2560	1995	2192			
2570	2005	2200			
2580	2016	2208			
2590	2026	2216			

FIGURE 20.—Airplane Weight and Balance Tables.

Chapter 6: Weight and Balance

Figures 19 and 20 are presented on pages 96 and 97.

11.
0188. (Refer to figures 19 and 20.) Which is the maximum amount of baggage that can be carried when the airplane is loaded as follows?

Front seat occupants	387 lb
Rear seat cargo	293 lb
Fuel	35 gal

A— 45 pounds.
B— 63 pounds.
C— 220 pounds.
D— 255 pounds.

Answer (A) is correct (0188). *(PHAK Chap IV)*

The maximum allowable weight on the Moment Limits vs Weight chart (Fig. 20) is 2,950 lb. The total of the given weights is 2,905 lb (including the empty weight of the airplane at 2,015 lb and the fuel at 6 lb/gal), so baggage cannot weigh more than 45 lb.

It is still necessary to compute total moments to verify that the position of these weights does not throw the plane out of CG limits. Prepare the weights and moments schedule as explained in the answer to question 10.

The total moment of 2,460 lies safely between the moment limits of 2,422 and 2,499 on Fig. 20, at the maximum weight, so this airplane can carry as much as 45 lb of baggage when loaded in this manner.

	Weight	Moment/100 lb inches
Empty weight with oil	2,015	1,554
Front seat	387	330
Rear seat	293	355
Fuel, main (35 gal x 6 lb)	210	158
Baggage	45	63
	2,950	2,460

12.
0189. (Refer to figures 19 and 20.) Calculate the weight and balance and determine if the CG and the weight of the airplane are within limits.

Front seat occupants	350 lb
Rear seat cargo	325 lb
Baggage	27 lb
Fuel	35 gal

A— 81.7, out of limits forward.
B— 83.4, within limits.
C— 84.1, within limits.
D— 84.8, out of limits aft.

Answer (B) is correct (0189). *(PHAK Chap IV)*

Total weight, total moment, and CG must all be calculated. As in most weight and balance problems, you should begin by setting up the schedule as shown below. The weights are given and the moments are determined as described in the answer to question 10.

Next, go to the Moment Limits vs Weight chart (Fig. 20), and note that the maximum weight allowed is 2,950, which means that this airplane is 23 lb under maximum weight. At a total moment of 2,441, it is also within the CG limits (2,399 to 2,483) at that weight.

Finally, compute the CG. Recall that Figure 19 gives moment per 100 inches. The total moment is therefore 244,100 (2,441 x 100). The CG is 244,100 ÷ 2,927 = 83.4.

	Weight	Moment/100 lb inches
Empty weight with oil	2,015	1,554
Front seat	350	298
Rear seat	325	393
Fuel, main (35 gal x 6 lb)	210	158
Baggage	27	38
	2,927	2,441

Chapter 6: Weight and Balance

Figures 19 and 20 are presented on pages 96 and 97.

13.
0191. (Refer to figures 19 and 20.) Determine if the airplane weight and balance are within limits.

Front seat occupants .... 415 lb
Rear seat cargo .... 110 lb
Fuel, main and aux. tanks .... 63 gal
Baggage .... 32 lb

A— 19 pounds overweight, CG within limits.
B— 19 pounds overweight, CG out of limits forward.
C— Weight within limits, CG out of limits.
D— Weight and balance within limits.

Answer (C) is correct (0191). *(PHAK Chap IV)*

Both the weight and the total moment must be calculated. Begin by setting up the schedule shown below. The fuel must be separated into main and auxiliary tanks, but weights and moments for both tanks are provided in Fig. 19.

In computing the moment for the front seat, use 130 lb + 140 lb + an interpolation for 145 lb. These moments are 110 + 119 + 124 = 353. The rear seat moment must be interpolated back from the 120 lb moment on the chart. Since the moments increase an average of 12, estimating a moment of 133 for 110 lb will not be too far off.

The last step is to go to the Moment Limits vs Weight chart (Figure 20). The maximum weight allowed is 2,950, which means that the airplane weight is within the limits.

However, the CG is out of limits because the minimum moment for a weight of 2,950 lb is 2,422.

	Weight	Moment/100 lb inches
Empty weight with oil	2,015	1,554
Front seat	415	353
Rear seat	110	133
Fuel, main	264	198
Fuel, auxiliary	114	107
Baggage	32	45
	2,950	2,390

14.
0192. (Refer to figures 19 and 20.) Which action can adjust the airplane's weight to maximum gross weight and the CG within limits for takeoff?

Front seat occupants .... 425 lb
Rear seat cargo .... 300 lb
Fuel, main tanks .... 44 gal

A— Drain 9 gallons of fuel.
B— Drain 12 gallons of fuel.
C— Transfer 12 gallons of fuel from the main tanks to the auxiliary tanks.
D— Transfer 19 gallons of fuel from the main tanks to the auxiliary tanks.

Answer (A) is correct (0192). *(PHAK Chap IV)*

First, determine the total weight to see how much must be reduced. As shown below, this original weight is 3,004 lb. Fig. 20 shows the maximum weight as 2,950 lb. Thus, you must adjust the total weight by removing 54 lb (3,004 - 2,950). Since fuel weighs 6 lb/gal, you must drain at least 9 gallons.

You know that Answers (C) and (D) are incorrect because transferring fuel to auxiliary tanks will only affect the moment, not the total weight.

To check for CG, recompute the total moment using a new fuel moment of 158 (from the chart) for 210 lb. The plane now weighs 2,950 lb with a total moment of 2,437, which falls within the moment limits on Figure 20.

	Original Weight	Adj. Weight	Moment/100 lb. inches
Empty weight w/oil	2,015	2,015	1,554
Front seat	425	425	362
Rear seat	300	300	363
Fuel	264	210	158
	3,004	2,950	2,437

Chapter 6: Weight and Balance

Figures 19 and 20 are presented on pages 96 and 97.

15.
0190. (Refer to figures 19 and 20.) Upon landing, a box in the front seat (180 pounds) is unloaded. A rear passenger (204 pounds) moves to the front passenger position. What effect does this have on the CG if the airplane weighed 2,690 pounds and the MOM/100 was 2,260 just prior to the passenger transfer?

A— The CG moves forward approximately 0.1 inch.
B— The CG moves forward approximately 2.4 inches.
C— The CG moves forward approximately 3 inches.
D— The weight changes, but the CG is not affected.

Answer (C) is correct (0190). *(PHAK Chap IV)*
The requirement is the effect of a change in loading. Look at Fig. 19 for occupants. Losing the 180-lb box from the front seat reduces the MOM/100 by 153. Moving the 204-lb passenger from the rear seat to the front reduces the MOM/100 by about 74 (247 - 173).
The total moment reduction is thus about 227 (153 + 74). As calculated below, the CG moves forward from 84.01 to 81.00 inches.

$$\text{Old CG} = \frac{226{,}000 \text{ lb-in}}{2{,}690 \text{ lb}} = 84.01 \text{ in}$$

$$\text{New CG} = \frac{203{,}300 \text{ lb-in}}{2{,}510 \text{ lb}} = 81.00 \text{ in}$$

16.
0193. (Refer to figures 19 and 20.) What effect does a 35-gallon fuel burn have on the weight and balance if the airplane weighed 2,890 pounds and the MOM/100 was 2,452 at takeoff?

A— Weight is reduced by 210 pounds and the CG is aft of limits.
B— Weight is reduced by 210 pounds and the CG is unaffected.
C— Weight is reduced to 2,680 pounds and the CG moves forward.
D— Weight is reduced to 2,855 pounds and the CG moves aft.

Answer (A) is correct (0193). *(PHAK Chap IV)*
The effect of a 35-gal fuel burn on weight balance is required. Burning 35 gal of fuel will reduce weight by 210 lb and moment by 158.
At 2,680 lb (2,890 - 210), the 2,294 MOM/100 (2,452 - 158) is above the maximum moment of 2,287; i.e., CG is aft of limits.

CHAPTER SEVEN
WEATHER

Causes of Weather	(2 questions)	102, 110
Fronts	(3 questions)	102, 110
Thunderstorms	(10 questions)	102, 111
Icing	(3 questions)	103, 113
Wind Shear	(2 questions)	103, 114
Temperature/Dewpoint and Fog	(9 questions)	103, 115
Clouds	(7 questions)	103, 117
Stability of Air Masses	(10 questions)	104, 119
Temperature Inversions	(4 questions)	104, 121
Weather Briefings	(2 questions)	105, 122
Surface Weather Reports	(5 questions)	105, 122
Area Forecasts	(3 questions)	106, 124
Terminal Forecasts	(6 questions)	106, 124
Weather Depiction Charts	(5 questions)	107, 126
Radar Summary Charts	(2 questions)	108, 128
En Route Flight Advisory Service	(2 questions)	108, 129
Low-Level Prognostic Charts	(3 questions)	108, 131
Transcribed Weather Broadcasts (TWEBs)	(5 questions)	108, 133
Winds and Temperatures Aloft Forecasts	(3 questions)	109, 134
AIRMETs and SIGMETs	(3 questions)	109, 135

This chapter contains outlines of major concepts tested, all FAA test questions and answers regarding weather, and an explanation of each answer. The subtopics or modules within this chapter are listed above, followed in parentheses by the number of questions from the FAA written test pertaining to that particular module. The two numbers following the parentheses are the page numbers on which the outline and questions begin for that module.

CAUTION: Recall that the sole purpose of this book is to expedite your passing the FAA written test for the recreational pilot certificate. Accordingly, all extraneous material (i.e., topics or regulations not directly tested on the FAA written test) is omitted, even though much more information and knowledge are necessary to fly safely. This additional material is presented in *RECREATIONAL PILOT FLIGHT MANEUVERS* and *PRIVATE PILOT HANDBOOK*, available from Aviation Publications, Inc. The weather chapter in *Private Pilot Handbook*, "Understanding Weather," is highly recommended to you to help you understand aviation weather. See pages 208 to 210 for more information and an order form.

CAUSES OF WEATHER (2 questions)

1. Unequal heating of the Earth's surface causes differences in pressure and altimeter settings.
2. The Coriolis force deflects winds to the right in the Northern Hemisphere. It is caused by the Earth's rotation.
   a. The Coriolis force is less at the surface due to less strong wind.
   b. Wind is less strong at the surface due to friction between wind and the Earth's surface.

FRONTS (3 questions)

1. A *front* is the zone of transition between two air masses of different density, e.g., the area separating a high pressure system and a low pressure system.
2. There is always a change in wind direction when flying across a front.
3. Sea breezes are caused by cooler and denser air moving inland off of the water. Land absorbs and radiates heat faster than water, so the air heats up and rises. Currents push the air over the water where it cools and descends, starting the process over again.
4. Convective circulation patterns associated with sea breezes are caused by land absorbing and radiating heat faster than the water.

THUNDERSTORMS (10 questions)

1. Thunderstorms are developed from cumulonimbus clouds. They form when there is
   a. Sufficient water vapor,
   b. An unstable air, and
   c. An initial upward boost to start the process.
2. Thunderstorms have three phases in their life cycle.
   a. **Cumulus:** The building stage of a thunderstorm when there are continuous updrafts.
   b. **Mature:** The time of greatest intensity when there are both updrafts and downdrafts (causing severe wind shear and turbulence).
      1) The commencing of rain on the Earth's surface indicates the *mature stage* of a thunderstorm.
   c. **Dissipating:** When there are only downdrafts; i.e., the storm is raining itself out.
3. A *thunderstorm*, by definition, has lightning because that is what causes thunder.
4. Thunderstorms produce *wind shear turbulence*, a hazardous and invisible phenomenon, particularly for airplanes landing and taking off.
   a. Hazardous wind shear near the ground can also be present during periods of strong temperature inversion.
5. The most severe thunderstorm conditions (heavy hail, destructive winds, tornadoes, etc.) are generally associated with squall line thunderstorms.
   a. A *squall line* is a <u>nonfrontal</u> narrow band of active thunderstorms that usually develops <u>ahead of a cold front</u>.
6. Embedded thunderstorms are obscured from visual location (i.e., pilots cannot see them) because they occur in very cloudy conditions.

Chapter 7: Weather   103

ICING (3 questions)

1. Structural icing requires two conditions:
    a. Flight through visible moisture, and
    b. The temperature at freezing or below.
2. Freezing rain usually causes the greatest accumulation of structural ice.
3. Ice pellets are caused when rain droplets freeze at a higher altitude, i.e., *freezing rain exists above*.

WIND SHEAR (2 questions)

1. *Wind shear* can occur at any level and be horizontal and/or vertical, i.e., **whenever adjacent air is flowing in different directions or speeds**.
2. Expect wind shear in a temperature inversion whenever wind speed at 2,000 to 4,000 ft AGL is 25 kts or more.

TEMPERATURE/DEWPOINT AND FOG (9 questions)

1. When the air temperature is within 4° of the dewpoint temperature and the spread is decreasing, you should expect fog and/or low clouds.
    a. *Dewpoint* is the temperature at which the air will have 100% humidity, i.e., be saturated.
    b. Thus, air temperature determines how much water vapor can be held by the air.
    c. Frost forms when both the collecting surface is below the dewpoint of the adjacent air AND the dewpoint is below freezing. *Frost* is the direct sublimation of water vapor to ice crystals.
2. Water vapor becomes visible as it condenses into clouds, fog, or dew, or sublimates into frost.
3. *Evaporation* -- conversion of liquid to water vapor.
    a. *Sublimation* -- conversion of ice to water vapor.
4. *Radiation fog* (shallow fog) is most likely to occur when there is a clear sky, little or no wind, and a small temperature/dewpoint spread.
5. *Advection fog* forms as a result of moist air condensing as it moves over a cooler surface.
6. Advection fog and upslope fog depend on a wind.

CLOUDS (7 questions)

1. Clouds are divided into four families based on their height:
    a. High clouds,
    b. Middle clouds,
    c. Low clouds, and
    d. Clouds with extensive vertical development.
2. *Nimbus* means rain cloud.
3. **Lifting action, unstable air, and moisture are the ingredients for the formation of cumulonimbus clouds.**

4. The greatest turbulence is in cumulonimbus clouds.

5. Towering cumulus are early stages of cumulonimbus, and usually indicate convective turbulence.

6. The altitude of the bases of cumulus clouds can be estimated based on an assumed convergence of rising air in a convective current, cooling at 5.4°/1,000 ft and the decrease in dewpoint of 1°/1,000 ft.

    a. Thus, the temperature/dewpoint spread divided by 4.4° equals the bases of cumulus clouds in thousands of feet.

    b. EXAMPLE. Surface dewpoint 56°, surface temperature 69° results in an estimate of cumulus cloud bases at 3,000 ft AGL: 69° - 56° = 13° temperature/dewpoint spread; 13° ÷ 4.4°/1,000 ft = approximately 3,000 ft.

STABILITY OF AIR MASSES (10 questions)

1. Stable air characteristics:

    a. Stratiform clouds
    b. Smooth air
    c. Fog
    d. Continuous precipitation
    e. Fair to poor visibility in haze and smoke

2. Moist, stable air moving up a mountain slope produces stratus-type clouds as it cools.

3. Unstable air characteristics:

    a. Cumuliform clouds
    b. Turbulent air
    c. Good visibility
    d. Showery precipitation

4. When air is warmed from below, it rises and causes instability.

5. Turbulence and clouds with vertical development result when unstable air rises.

6. The *lapse rate* is the decrease in temperature with increase in altitude. As the lapse rate increases (i.e., air cools more with increases in altitude), air is more unstable.

    a. The lapse rate can be used to determine the stability of air masses.

TEMPERATURE INVERSIONS (4 questions)

1. Normally, temperature decreases as altitude increases. A *temperature inversion* occurs when temperature increases as altitude increases.

2. Temperature inversions usually result in a stable layer of air.

3. A temperature inversion often develops near the ground on clear, cool nights when the wind is light.

    a. It is caused by terrestrial radiation.

4. Smooth air with restricted visibility is usually found beneath a low level temperature inversion.

Chapter 7: Weather

WEATHER BRIEFINGS (2 questions)

1. When requesting a telephone weather briefing, you should identify
    a. Yourself as a recreational pilot, i.e., you intend to fly VFR only, and
    b. Your proposed departure time and area of flight.

SURFACE WEATHER REPORTS (5 questions)

1. *Surface aviation weather reports* are actual weather observations made once each hour at each reporting station.
    a. They are available by teletype and by computer in all flight service stations and coded by the letters "SA."
        1) "RS" means "record special" weather report.
    b. The reports are also referred to as sequence reports.

2. Three-letter station identifiers are listed in the column on the left side of the report and are followed by
    a. The designation or type of report, i.e., SA.
    b. The time the weather observations were made using Greenwich Mean Time.
    c. The weather observation at that station.

3. The weather is presented as
    a. The basic sky conditions and ceilings.
        1) *Ceiling* is the height above the earth's surface of the lowest layer of clouds that is reported as broken, overcast, or obscured (except those classified as thin, partial, or scattered).
    b. Visibility.
    c. Weather and obstructions to vision.
    d. Sea level pressure (millibars).
    e. Temperature and dewpoint.
    f. Wind direction, speed, and character.
    g. Altimeter setting (inches of Hg).
    h. Remarks.

4. EXAMPLE. SLC SA 1251 E110 OVC 30 079/53/28/1916G24/981/RF2RB12
    a. **SLC** is Salt Lake City.
    b. **SA** is sequence report (hourly weather observation).
    c. **1251** is the time of the weather observations.
    d. **E110 OVC** is estimated ceiling at 11,000 ft AGL which is overcast.
        1) "-X" means partial sky obscuration.
    e. Visibility is **30** statute miles.
    f. Pressure is **1**007.9 millibars.
    g. Temperature is **53°** F.

106　Chapter 7: Weather

    h. Dewpoint is **28°** F.

    i. Wind is from **190°** at **16** kts, **G**usting to **24** kts.

    j. Altimeter setting is **29.81**.

    k. Remarks: **R**ain and **F**og obscuring two-tenths (2/10) of the sky. **R**ain **B**egan **12** minutes past the hour.

        1) "Rt" means heavy rain.

AREA FORECASTS (3 questions)

1. *Area forecasts (FA)* are for several states and/or portions of states. Issued every 12 hrs and valid for a period of 30 hrs, they are
    a. Expected weather for 18 hrs.
    b. An additional 12-hr categorical outlook.

2. The order of topics in area forecasts, each being on a separate line or in a separate paragraph, is
    a. Heading
    b. Forecast area
    c. Height statement (i.e., feet MSL or AGL)
    d. Flight precaution statement
    e. Synopsis
    f. Significant clouds and weather, plus categorical outlook
        1) A summary of cloudiness and weather significant to flight operations broken down by states or other geographical areas.
    g. Icing and freezing level
    h. Turbulence
    i. AIRMETs and SIGMETs

TERMINAL FORECASTS (6 questions)

1. *Terminal forecasts* are weather forecasts for large airports throughout the country.

2. Forecasts are issued three times a day for the next 24-hr period.
    a. The last 6 hours are an "outlook" rather than a forecast.
        1) WIND means wind in excess of 25 kts. CIG means ceilings. TRW means thundershowers.

3. Terminal forecasts are identified by the letters "FT."

4. The components of the forecast for each station are
    a. Station identifier.
    b. 6-digit number: first 2 digits indicate the date; second 2 digits indicate the time Zulu the forecast begins today; third 2 digits indicate the time Zulu the forecast ends tomorrow.
    c. *Zulu* is Greenwich Mean Time (also abbreviated GMT).
    d. The first 6-hr forecast begins with cloud and wind forecasts.
    e. Each subsequent 6-hr forecast is preceded by the time Zulu it begins.

Chapter 7: Weather

5. EXAMPLE. GNV 181010 100 SCT 250 -BKN 1615. 18Z C80 BKN 1815. 00Z C50 BKN 3215. 04Z MVFR CIG.

    a. **GNV** is Gainesville.

    b. **181010.** The forecast is made on the 18th day of the month and is valid from **1000Z** on the 18th to **1000Z** on the 19th.

    c. **100 SCT 250 -BKN 1615.** This means clouds at **10,000 scattered, 25,000** thin **broken**, with surface wind from **160°** at **15** kts.

    d. **18Z C80 BKN 1815.** This means from **1800Z**, the ceiling is forecast to be **8,000** broken with surface wind from 180° at 15 kts.

    e. **00Z C50 BKN 3215.** This means from **0000Z**, the ceiling is forecast to be **5,000 broken** with surface wind from **320°** at **15** kts.

    f. **04Z MVFR CIG.** This means from **0400Z**, marginal **VFR** is forecast due to ceilings.

WEATHER DEPICTION CHARTS (5 questions)

1. A *weather depiction chart* is an outline of the United States depicting sky conditions at a time stated on the chart.

    a. Selected reporting stations are marked with a little circle.
        1) If the sky is clear, the circle is open; if overcast, the circle is solid; if scattered, the circle is ¼ solid; if broken, the circle is ¾ solid. If the sky is obscured, there is an "X" in the circle.
        2) The height of clouds is expressed in hundreds of feet, e.g., 120 means 12,000 ft.

2. Areas with ceilings below 1,000 ft and/or visibility less than 3 miles, i.e., below VFR, are bracketed with solid black lines.

    a. Visibility is indicated on the circle, e.g., 2 stands for 2 miles visibility.

    b. Areas of marginal VFR with ceilings of 1,000 to 3,000 ft and/or visibility at 3 to 5 miles are bracketed by scalloped lines, i.e., lines made up of continuous rounded projections.

    c. Ceilings greater than 3,000 ft and visibility greater than 5 miles are not indicated on weather depiction charts.

3. Significant weather is indicated by the following symbols:

△ ICE PELLETS	▽ SHOWER	= FOG OR GROUND FOG
✳ SNOW	⚡ THUNDERSTORM	∞ HAZE
•• RAIN		∿ SMOKE
⚡ FREEZING DRIZZLE	▲ CLOUDS TOPPING RIDGES	﹐﹐ DRIZZLE
⚡ FREEZING RAIN		

4. The weather depiction chart quickly shows pilots where weather conditions reported are above or below VFR minimums.

5. Stationary fronts are indicated by rounded scallops on one side of the frontal line and pointed scallops on the other side.

RADAR SUMMARY CHARTS (2 questions)

1. *Radar summary charts* graphically display a collection of radar reports concerning the intensity and movement of precipitation, e.g., squall lines, specific thunderstorm cells, and other areas of hazardous precipitation.

EN ROUTE FLIGHT ADVISORY SERVICE (2 questions)

1. *En Route Flight Advisory Service (EFAS)* provides timely and meaningful weather advisories to en route aircraft on 122.0 MHz. The broadcast is called "Flight Watch."
    a. Generally available from 6 a.m. to 10 p.m. local time.
2. EFAS provides a continual exchange of information on winds, turbulence, visibility, icing, etc. to pilots en route.

LOW-LEVEL PROGNOSTIC CHARTS (3 questions)

1. Low-level prognostic charts contain four charts (panels).
    a. The two upper panels forecast significant weather from the surface up to 24,000 ft: one for 12 hrs and the other for 24 hrs from the time of issuance.
    b. The two lower panels forecast surface conditions: one for 12 hrs and the other for 24 hrs from time of issuance.
2. The top panels show
    a. Ceilings less than 1,000 ft, visibility less than 3 miles (IFR) by a solid line around the area;
    b. Marginal VFR by a scalloped line around the area;
    c. Moderate or greater turbulence by a broken line around the area; and
    d. Freezing levels, given by a dashed line corresponding to the height of the freezing level.
3. The bottom panels show the location of
    a. Highs, lows, fronts, and
    b. Other areas of significant weather.
4. See the weather symbols under Weather Depiction Charts on page 107.
    a. Two symbols indicate continuous and one symbol indicates intermittent.

TRANSCRIBED WEATHER BROADCASTS (TWEBs) (5 questions)

1. TWEBs are continuous recordings of meterological and aeronautical information broadcast on certain NDB and VOR facilities,
    a. Generally based on specific routes of flight.

Chapter 7: Weather

WINDS AND TEMPERATURES ALOFT FORECASTS (3 questions)

1. Forecast winds and temperatures at specified altitudes for specific locations in the United States.
2. A four-digit group (used when temperatures are not forecast) shows wind direction with reference to true north and the wind speed in knots.
   a. The first two digits indicate the wind direction after you add a zero.
   b. The next two digits indicate the windspeed.
3. A six-digit group includes the forecast temperature aloft.
   a. The last two digits indicate the temperature in degrees Celsius.
   b. Plus or minus is indicated before the temperature, except at higher altitudes (above 24,000 ft MSL) where it is always below freezing.
4. When the wind speed is less than 5 kts, the forecast is coded 9900, which means that the wind is light and variable.
5. An example forecast is provided on page 130 for questions 76 and 77.

AIRMETS AND SIGMETS (3 questions)

1. SIGMETs and AIRMETs are issued to notify pilots en route of the possibility of encountering hazardous flying conditions.
2. SIGMET advisories include weather phenomena that are potentially hazardous to all aircraft.
   a. Convective SIGMETs include
      1) Tornadoes,
      2) Lines of thunderstorms,
      3) Embedded thunderstorms,
      4) Thunderstorm areas greater than or equal to thunderstorm intensity level 4 with an area coverage of 40% or more, and
      5) Hail greater than or equal to ¾ inch in diameter.
   b. Other SIGMETs include
      1) Severe and extreme turbulence,
      2) Severe icing,
      3) Widespread duststorms, sandstorms, or volcanic ash lowering visibilities to less than 3 miles.
3. AIRMETs apply to light aircraft to notify of
   a. Moderate icing,
   b. Moderate turbulence,
   c. Visibility less than 3 miles or ceilings less than 1,000 ft,
   d. Sustained winds of 30 kts or more at the surface, and
   e. Extensive mountain obscurement.

Chapter 7: Weather

> **QUESTIONS AND ANSWER EXPLANATIONS**
>
> All of the FAA questions from the written test for the Recreational Pilot certificate relating to airplanes and weather and the material outlined previously are reproduced below in the same modules as the previous outlines. To the immediate right of each question are the correct answer and answer explanation. You should cover these answers and answer explanations while responding to the questions. Refer to the general discussion in Chapter 1 on how to take the examination.
>
> Remember that the questions from the FAA's Written Test Book have been reordered by topic, and the topics have been organized into a meaningful sequence. Accordingly, the first line of the answer explanation gives the FAA question number and the citation of the authoritative source for the answer.

CAUSES OF WEATHER

1.
0400. What causes variations in altimeter settings between weather reporting points?

A— Unequal heating of the Earth's surface.
B— Variation of terrain elevation creating barriers to the movement of an air mass.
C— Coriolis force reacting with friction.
D— Friction of the air with the Earth's surface.

Answer (A) is correct (0400). *(AvW Chap 1)*

Unequal heating of the Earth's surface causes differences in air pressure, which is reflected in differences in altimeter settings between weather reporting points.

Answer (B) is incorrect because variation of terrain elevation creates barriers to air mass movement which has an effect on wind. It may cause turbulence near the surface. Answer (C) is incorrect because Coriolis force deflects wind to the right in the Northern Hemisphere. Near the surface the Coriolis force is decreased by friction caused by the Earth's surface. Answer (D) is incorrect because friction between wind and the terrain slows wind.

2.
0402. Winds at 5,000 feet AGL on a particular flight are southwesterly while most of the surface winds are southerly. This difference in direction is primarily due to

A— a stronger pressure gradient at higher altitudes.
B— friction between the wind and the surface.
C— stronger Coriolis force at the surface.
D— the influence of pressure systems at the lower altitudes.

Answer (B) is correct (0402). *(AvW Chap 4)*

Winds aloft at 5,000 ft are largely affected by Coriolis force, which deflects wind to the right. But at the surface, the winds will be more southerly (they were southwesterly aloft) because Coriolis force has less effect at the surface where the wind-speed is slower. The windspeed is slower at the surface due to the friction between the wind and the surface.

Answers (A) and (D) are incorrect because the pressure differentials are approximately uniform throughout the altitudes. Answer (C) is incorrect because the Coriolis force at the surface is weaker with slower windspeed.

FRONTS

3.
0433. The boundary between two different air masses is referred to as a

A— foehn gap.
B— frontolysis.
C— frontogenesis.
D— front.

Answer (D) is correct (0433). *(AvW Chap 8)*

A front is the boundary (or zone of transition) between two different air masses.

Answer (A) is incorrect because a *foehn* is a warm, dry, downslope wind. Answer (B) is incorrect because *frontolysis* is the dissipation of a front. Answer (C) is incorrect because *frontogenesis* is the initial formation of a front or frontal zone.

Chapter 7: Weather

4.
0432. One weather phenomenon which will always occur when flying across a front is

A— a change in the wind.
B— a large precipitation area, if the frontal surface is steep.
C— a large temperature change, especially at high altitudes.
D— the presence of clouds, either ahead of or behind the front.

Answer (A) is correct (0432). *(AvW Chap 8)*
The definition of a front is the zone of transition between two air masses of different air pressure or density, e.g., the area separating high and low pressure systems. Due to the difference in changes in pressure systems, there will be a change in wind.
Answer (B) is incorrect because the precipitation area will be small if the frontal surface is steep. If it is shallow, it will be larger. Answer (C) is incorrect because the difference in temperature is greater at low altitudes. Answer (D) is incorrect because, if there is not much moisture available, there will be relatively little cloud cover.

5.
0464. Convective circulation patterns associated with sea breezes are caused by

A— warm and less dense air moving inland from over the water, causing it to rise.
B— water absorbing and radiating heat faster than the land.
C— land absorbing and radiating heat faster than the water.
D— cool and less dense air moving inland from over the water, causing it to rise.

Answer (C) is correct (0464). *(PHAK Chap V)*
Sea breezes are caused by cool and more dense air moving inland off the water. Once over the warmer land, the air heats up and rises. Thus the cooler, more dense air from the sea forces the warmer air up. Currents push the hot air over the water where it cools and descends, starting the cycle over again. This process is caused by land heating up and cooling off faster than water.
Answer (A) is incorrect because the air over the water is cooler (not warmer) and more (not less) dense. Answer (B) is incorrect because water absorbs and radiates heat slower (not faster) than land. Answer (D) is incorrect because the cool air moving inland is more (not less) dense.

THUNDERSTORMS

6.
0443. What conditions are necessary for the formation of thunderstorms?

A— Lifting force, high humidity, and unstable conditions.
B— High humidity, high temperature, and cumulus clouds.
C— Low pressure, high humidity, and cumulus clouds.
D— Lifting force, high temperature, and unstable conditions.

Answer (A) is correct (0443). *(AvW Chap 11)*
Thunderstorms form when there is sufficient water vapor, unstable air, and an initial upward boost to start the process.
Answers (B) and (D) are incorrect because a high temperature is not necessary; e.g., thunderstorms can occur in the arctic. Answer (C) is incorrect because low pressure is not required.

7.
0455. What feature is normally associated with the cumulus stage of a thunderstorm?

A— Roll cloud.
B— Continuous updraft.
C— Frequent lightning.
D— Beginning of rain at the surface.

Answer (B) is correct (0455). *(AvW Chap 11)*
The cumulus stage of a thunderstorm has continuous updrafts which build the storm. The water droplets are carried up until they become too heavy. Once they begin falling and creating downdrafts, the storm changes from the cumulus to the mature stage.
Answer (A) is incorrect because the roll cloud is the cloud on the ground which is formed by the downrushing cold air pushing out from underneath the bottom of the thunderstorm. Answer (C) is incorrect because frequent lightning is associated with the mature stage where there is a considerable amount of wind shear and static electricity. Answer (D) is incorrect because the beginning of rain at the surface indicates the beginning of the mature stage, which follows the cumulus stage.

8.

0456. Which weather phenomenon signals the beginning of the mature stage of a thunderstorm?

A— The appearance of an anvil top.
B— The start of rain at the surface.
C— Growth rate of cloud is maximum.
D— Strong turbulence in the cloud.

Answer (B) is correct (0456). *(AvW Chap 11)*

The mature stage of a thunderstorm begins when rain begins falling. This means that the downdrafts are occurring sufficiently to carry water all the way through the thunderstorm.

Answer (A) is incorrect because the anvil top generally appears during, but not necessarily at the beginning, of the mature stage. Answers (C) and (D) are incorrect because they are further into the mature stage of a thunderstorm, not at the beginning.

9.

0445. Thunderstorms reach their greatest intensity during the

A— updraft stage.
B— mature stage.
C— downdraft stage.
D— cumulus stage.

Answer (B) is correct (0445). *(AvW Chap 11)*

Thunderstorms reach their greatest intensity during the mature stage, where updrafts and downdrafts cause a high level of wind shear.

Answers (A) and (D) are incorrect because the cumulus stage, which consists mainly of building, is not the most intense until the downdrafts begin occurring as a result of cold water drops falling. Answer (C) is incorrect because the downdraft stage is known as the dissipating stage, which is when the thunderstorm rains itself out.

10.

0444. During the life cycle of a thunderstorm, which stage is characterized predominately by downdrafts?

A— Cumulus.
B— Dissipating.
C— Mature.
D— Anvil.

Answer (B) is correct (0444). *(AvW Chap 11)*

Thunderstorms have three life cycles: cumulus, mature, and dissipating. It is in the dissipating stage that the storm is characterized by downdrafts as the storm rains itself out.

Answer (A) is incorrect because cumulus is the building stage when there are updrafts. Answer (C) is incorrect because the mature stage is when there are both updrafts and downdrafts, which creates tremendous wind shears. Answer (D) is incorrect because the anvil is the top of the cloud which points in the direction of the storm's movement.

11.

0458. Which weather phenomenon is always associated with a thunderstorm?

A— Lightning.
B— Heavy rain showers.
C— Supercooled raindrops.
D— Hail.

Answer (A) is correct (0458). *(AvW Chap 11)*

A thunderstorm, by definition, has lightning, because lightning causes the thunder.

Answer (B) is incorrect because, while heavy rain showers usually occur, hail may occur instead. Answers (C) and (D) are incorrect because supercooled raindrops and hail may not occur if the lifting process does not extend above the freezing level.

12.

0436. Hazardous wind shear is commonly encountered near the ground

A— near thunderstorms and during periods when the wind velocity is stronger than 35 knots.
B— during periods when the wind velocity is stronger than 35 knots and near mountain valleys.
C— during periods of strong temperature inversion and near thunderstorms.
D— near mountain valleys and on the windward side of a hill or mountain.

Answer (C) is correct (0436). *(AvW Chap 9)*

Wind shear near the ground is produced with a low-level temperature inversion or in a frontal zone near thunderstorms.

Answers (A) and (B) are incorrect because turbulence at all altitudes is usually prevalent when wind velocity is stronger than 35 kts. Answer (D) is incorrect because, on the windward side of a hill or mountain, the winds are rising, assisting the pilot. It is the wind on the leeward side that is hazardous.

Chapter 7: Weather 113

13.
0459. A nonfrontal, narrow band of active thunderstorms, that often develop ahead of a cold front, is known as

A— an occlusion.
B— a prefrontal system.
C— a squall line.
D— a shear line.

Answer (C) is correct (0459). *(AvW Chap 11)*
A nonfrontal narrow band of active thunderstorms that develops ahead of a cold front is known as a squall line.
Answer (A) is incorrect because an occlusion is a composition of two fronts such as when a cold front overtakes a warm front or a stationary front. Answer (B) is incorrect because a prefrontal system is weather that precedes a front. Answer (D) is incorrect because a shear line, if it is not a non-sense concept, is the point at which two wind systems are going in different directions, i.e., that which causes wind shear.

14.
0457. Thunderstorms which generally produce the most intense hazard to aircraft are

A— air mass thunderstorms.
B— steady-state thunderstorms.
C— warm front thunderstorms.
D— squall line thunderstorms.

Answer (D) is correct (0457). *(AvW Chap 11)*
A squall line is a nonfrontal narrow band of active thunderstorms. It often contains severe, steady-state thunderstorms and presents the single most intense weather hazard to airplanes.
Answers (A), (B), and (C) are incorrect because, although each is hazardous to airplanes, squall line thunderstorms are generally much more severe.

15.
0435. What is indicated when a current SIGMET forecasts embedded thunderstorms?

A— Thunderstorms have been visually sighted.
B— Severe thunderstorms are embedded within a squall line.
C— Thunderstorms are dissipating and present no serious problem to IFR flight.
D— Thunderstorms are obscured by massive cloud layers and cannot be seen.

Answer (D) is correct (0435). *(AvW Chap 11)*
Embedded thunderstorms are thunderstorms within cloud formations that cannot be seen by pilots, which is why they are so dangerous.
Answer (A) is incorrect because thunderstorms that are visually sighted can be avoided (and are not embedded). Answer (B) is incorrect because thunderstorms in a squall line are highly visible. Answer (C) is incorrect because dissipating thunderstorms which present no serious problem will not be the topic of a SIGMET. SIGMET is an acronym for "significant meteorological information," which is significant to all aircraft, including air carriers. It covers severe and extreme turbulence, severe icing, and widespread dust or sandstorms.

ICING

16.
0439. One in-flight condition necessary for structural icing to form is

A— cumuliform clouds.
B— cirrostratus clouds.
C— stratiform clouds.
D— visible moisture.

Answer (D) is correct (0439). *(AvW Chap 10)*
Two conditions are necessary for structural icing while in flight. First, the airplane must be flying through visible moisture, such as rain or cloud droplets. Second, the temperature at the point where the moisture strikes the airplane must be at or below freezing.
Answers (A), (B), and (C) are incorrect because no special cloud formation is necessary for icing as long as visible moisture is present.

114 Chapter 7: Weather

17.
0440. In which environment is aircraft structural ice most likely to have the highest accumulation rate?

A— Cumulus clouds.
B— Cirrus clouds.
C— Stratus clouds.
D— Freezing rain.

Answer (D) is correct (0440). *(AvW Chap 10)*
Freezing rain usually causes the highest accumulation rate of structural icing because of the nature of the supercooled water striking the airplane.
Answers (A), (B), and (C) are incorrect because this condition of supercooled water may or may not exist in cumulus, cirrus, or stratus clouds, depending upon the precipitation level and the temperature.

18.
0407. The presence of ice pellets at the surface is evidence that

A— thunderstorms are in the area.
B— a cold front has passed.
C— freezing rain is at a higher altitude.
D— a stationary front is overhead.

Answer (C) is correct (0407). *(AvW Chap 10)*
Ice pellets form as a result of rain freezing at a higher altitude. Rain droplets cool as they fall from a warmer temperature through air with a temperature below freezing.
Answers (A) and (B) are incorrect because, while thunderstorms and a passing cold front may cause freezing rain, they are not the only cause of ice pellets. Answer (D) is incorrect because ice pellets are not commonly associated with a stationary front.

WIND SHEAR

19.
0437. Where does wind shear occur?

A— Only at higher altitudes, usually in the vicinity of jetstreams.
B— At any level, and it can exist in both a horizontal and vertical direction.
C— Primarily at lower altitudes in the vicinity of mountain waves.
D— Only in the vicinity of thunderstorms.

Answer (B) is correct (0437). *(AvW Chap 9)*
Wind shear occurs because of changes in wind direction and wind velocity, both horizontal and vertical. It may be present at any flight level.
Answers (A) and (C) are incorrect because wind shear occurs at all altitudes and can be vertical as well as horizontal. Answer (D) is incorrect because wind shear does occur in the vicinity of thunderstorms, but also whenever there are changes in wind direction or wind velocities.

20.
0438. A pilot can expect a wind shear zone in a temperature inversion, whenever the windspeed at 2,000 to 4,000 feet above the surface is at least

A— 5 knots.
B— 10 knots.
C— 15 knots.
D— 25 knots.

Answer (D) is correct (0438). *(AvW Chap 9)*
When taking off or landing in calm wind under clear skies within a few hours before or after sunset, prepare for a temperature inversion near the ground. Be relatively certain of a shear zone in the inversion if you know the wind is 25 kts or more at 2,000 to 4,000 ft. Allow a margin of airspeed above normal climb or approach speed to alleviate the danger of stall in the event of turbulence or sudden change in wind velocity.

Chapter 7: Weather

TEMPERATURE/DEWPOINT AND FOG

21.
0404. What is meant by the term dewpoint?

A— The temperature at which condensation and evaporation are equal.
B— The temperature at which dew will always form.
C— The temperature to which air must be cooled to become saturated.
D— The spread between actual temperature and the temperature during evaporation.

Answer (C) is correct (0404). *(AvW Chap 9)*
Dewpoint is the temperature to which air must be cooled to become saturated or have 100% humidity.
Answer (A) is incorrect because evaporation is the change from water to water vapor and is not directly related to the dewpoint. Answer (B) is incorrect because dew forms only when heat radiates from an object and its temperature lowers below the dewpoint. Answer (D) is incorrect because the spread in temperature is the latent heat absorbed in the change from water to water vapor.

22.
0405. The amount of water vapor which air can hold largely depends on

A— the dewpoint.
B— air temperature.
C— stability of air.
D— relative humidity.

Answer (B) is correct (0405). *(AvW Chap 9)*
Air temperature determines how much water vapor can be held by the air. Warm air can hold more water vapor than cold air.
Answer (A) is incorrect because the dewpoint is the temperature at which air with a given amount of moisture becomes totally saturated. Answer (C) is incorrect because air stability is the state of the atmosphere at which vertical distribution of temperature is such that air particles will resist displacement from their initial level. Answer (D) is incorrect because relative humidity is the ratio of the existing amount of water vapor in the air at a given temperature and the maximum amount that could exist at that temperature.

23.
0403. If the temperature/dewpoint spread is small and decreasing, and the temperature is 62 °F, what type weather is most likely to develop?

A— Freezing precipitation.
B— Thunderstorms.
C— Fog or low clouds.
D— Rain showers.

Answer (C) is correct (0403). *(AvW Chap 9)*
The difference between the air temperature and dewpoint temperature is the temperature/dewpoint spread. As the temperature/dewpoint spread decreases, fog or low clouds tend to develop.
Answer (A) is incorrect because there cannot be freezing precipitation if the temperature is 62°F. Answer (B) is incorrect because thunderstorms have to do with unstable lapse rates, not temperature/dewpoint spreads. Answer (D) is incorrect because rain showers will only occur if the air is saturated through thick layers aloft.

24.
0453. Which condition(s) result(s) in the formation of frost?

A— The freezing of dew.
B— The collecting surface's temperature is at or below freezing and small droplets of moisture fall on the collecting surface.
C— The temperature of the collecting surface is at or below the dewpoint of the adjacent air and the dewpoint is below freezing.
D— Small drops of moisture falling on the collecting surface when the surrounding air temperature is at or below freezing.

Answer (C) is correct (0453). *(AvW Chap 5)*
Frost forms when both the collecting surface is below the dewpoint of the adjacent air AND the dewpoint is below freezing. Frost is the direct sublimation of water vapor to ice crystals.
Answer (A) is incorrect because freezing of dew results in ice, not ice crystals (which is frost). Answers (B) and (D) are incorrect because, if moisture falls on a collecting surface that is below freezing, icing will occur.

Chapter 7: Weather

25.
0408. Clouds, fog, or dew will always form when

A— water vapor condenses.
B— water vapor is present.
C— relative humidity reaches or exceeds 100 percent.
D— the temperature and dewpoint are equal.

26.
0406. What are the processes by which moisture is added to unsaturated air?

A— Heating and sublimation.
B— Evaporation and sublimation.
C— Heating and condensation.
D— Supersaturation and evaporation.

27.
0461. What situation is most conducive to the formation of radiation fog?

A— Warm, moist air over low, flatland areas on clear, calm nights.
B— Moist, tropical air moving over cold, offshore water.
C— The movement of cold air over much warmer water.
D— Light wind moving warm, moist air upslope during the night.

28.
0463. In which situation is advection fog most likely to form?

A— A warm, moist air mass on the windward side of mountains.
B— An air mass moving inland from the coast in winter.
C— A light breeze blowing colder air out to sea.
D— Warm, moist air settling over a warmer surface under no-wind conditions.

Answer (A) is correct (0408). *(AvW Chap 9)*
 As water vapor condenses, it becomes visible as clouds, fog, or dew.
 Answer (B) is incorrect because water vapor is usually always present but does not form clouds, fog, or dew without condensation. Answers (C) and (D) are incorrect because even at 100% humidity (dewpoint equals actual temperature), water vapor may not condensate, e.g., sufficient condensation nuclei may not be present.

Answer (B) is correct (0406). *(AvW Chap 9)*
 Evaporation is the process of converting a liquid to water vapor, and sublimation is the process of converting solids to vapor, e.g., evaporation of ice.
 Answers (A) and (C) are incorrect because heating alone does not add or detract moisture. Answer (C) is also incorrect because condensation is the process of taking water out of air and converting it to liquid. Answer (D) is incorrect because supersaturation is a nonsense term in this context.

Answer (A) is correct (0461). *(AvW Chap 12)*
 Radiation fog is shallow fog of which ground fog is one form. It occurs under conditions of clear skies, little or no wind, and a small temperature/dewpoint spread. The fog forms almost exclusively at night or near daybreak as a result of terrestrial radiation cooling the ground and the ground cooling the air on contact with it.
 Answer (B) is incorrect because moist, tropical air moving over cold, offshore water causes advection fog, not radiation fog. Answer (C) is incorrect because movement of cold dry air over much warmer water results in steam fog. Answer (D) is incorrect because a light wind moving warm, moist air upslope during the night causes upslope fog.

Answer (B) is correct (0463). *(AvW Chap 12)*
 Advection fog forms when moist air moves over colder ground or water. It is most common in coastal areas.
 Answer (A) is incorrect because a warm, moist air mass on the windward side of mountains produces rain or upslope fog as it blows upward and cools. Answer (C) is incorrect because a light breeze blowing colder air out to sea causes steam fog. Answer (D) is incorrect because warm, moist air settling over a warmer surface under no-wind conditions can result in radiation fog if the warmer surface cools near the surface to below the dewpoint.

Chapter 7: Weather

29.
0462. What types of fog depend upon a wind in order to exist?

A— Radiation fog and ice fog.
B— Steam fog and downslope fog.
C— Precipitation-induced fog and ground fog.
D— Advection fog and upslope fog.

Answer (D) is correct (0462). *(AvW Chap 12)*
Upslope fog forms as a result of moist, stable air being cooled as it moves up a sloping terrain. Advection fog forms when moist air is blown over a cold surface, decreasing the moist air's temperature to its dewpoint. Thus, both upslope fog and advection fog depend on air moving from one area to another, i.e., depend on wind.

Answer (A) is incorrect because radiation fog is formed on nights when the sky is clear and wind is light or nonexistent. Ice fog describes tiny ice particals suspended in the atmosphere. Answer (B) is incorrect because steam fog forms when cold air moves over warm water or wet ground. Downslope fog is a nonsense term. Answer (C) is incorrect because precipitation fog forms when warm rain or drizzle falls through cool air. Ground fog is radiation fog.

CLOUDS

30.
0418. Clouds are divided into four families according to their

A— origin.
B— outward shape.
C— height range.
D— composition.

Answer (C) is correct (0418). *(AvW Chap 7)*
The four families of clouds are high clouds, middle clouds, low clouds, and clouds with extensive vertical development. Thus, they are based upon their height range.

Answer (A) is incorrect because the origin of all clouds is water vapor condensing on nuclei in the atmosphere. Answers (B) and (D) are incorrect because clouds' outward shape and composition have to do with their vertical stability.

31.
0417. The suffix nimbus, used in naming clouds, means

A— a cloud with extensive vertical development.
B— a rain cloud.
C— a middle cloud containing ice pellets.
D— an accumulation of clouds.

Answer (B) is correct (0417). *(AvW Chap 7)*
The suffix *-nimbus* or the prefix *nimbo-* means a rain cloud.

Answer (A) is incorrect because clouds with extensive vertical development are called either towering cumulus or cumulonimbus. Answer (C) is incorrect because a middle cloud containing ice pellets has the suffix *-alto*. Answer (D) is incorrect because an accumulation of clouds that covers the sky is classified as a ceiling.

32.
0411. The conditions necessary for the formation of cumulonimbus clouds are a lifting action and

A— unstable air containing an excess of condensation nuclei.
B— unstable, moist air.
C— either stable or unstable air.
D— stable, moist air.

Answer (B) is correct (0411). *(AvW Chap 11)*
Unstable moist air in addition to a lifting action, i.e., convective activity, are needed to form cumulonimbus clouds.

Answer (A) is incorrect because there must be moisture available to produce the clouds and rain; i.e., in a hot, dry dust storm, there would be no thunderstorm. Answers (C) and (D) are incorrect because the air must be unstable or there will be no lifting action.

33.
0419. Which clouds have the greatest turbulence?

A— Towering cumulus.
B— Cumulonimbus.
C— Nimbostratus.
D— Altocumulus castellanus.

Answer (B) is correct (0419). *(AvW Chap 7)*

The greatest turbulence occurs in cumulonimbus clouds, which are thunderstorm-type clouds.

Answer (A) is incorrect because towering cumulus are an earlier stage of cumulonimbus clouds. Answer (C) is incorrect because nimbostratus is a gray or dark, massive cloud layer diffused by more or less continuous rain or ice pellets. It is a middle cloud with very little turbulence but may pose serious icing problems. Answer (D) is incorrect because altocumulus castellanus clouds are middle-level convective clouds characterized by billowing tops and comparatively high bases. They indicate mid-level instability, but not nearly as extensive as cumulonimbus clouds.

34.
0420. Which cloud types would indicate convective turbulence?

A— Altocumulus standing lenticular clouds.
B— Nimbostratus clouds.
C— Towering cumulus clouds.
D— Cirrus clouds.

Answer (C) is correct (0420). *(AvW Chap 7)*

Towering cumulus clouds are an early stage of cumulonimbus clouds, or thunderstorms, which are based on convective turbulence, i.e., an unstable lapse rate.

Answer (A) is incorrect because lenticular clouds form on the crest of waves created by barriers (e.g., mountains) in wind flow, not convective turbulence. Answer (B) is incorrect because nimbostratus are gray or dark, massive clouds diffused by more or less continuous rain or ice pellets. Answer (D) is incorrect because cirrus clouds are high, thin, featherlike ice crystal clouds in patches and narrow bands which are not based on any convective activity.

35.
0416. At approximately what altitude above the surface would the pilot expect the base of cumuliform clouds if the surface air temperature is 82 °F and the dewpoint is 54 °F?

A— 5,000 feet AGL.
B— 6,000 feet AGL.
C— 7,000 feet AGL.
D— 9,000 feet AGL.

Answer (B) is correct (0416). *(AvW Chap 6)*

The height of cumuliform cloud bases can be estimated using surface temperature/dewpoint spread. Unsaturated air in a convective current cools at about 5.4°F/1,000 ft, and dewpoint decreases about 1°F/1,000 ft. In a convective current, temperature and dewpoint converge at about 4.4°F/1,000 ft. Thus, if the temperature/dewpoint spread is 28° (82° - 54°), divide 28° by 4.4°/1,000 ft to obtain 6.36 or about 6,000 ft AGL. This is a faulty question and the FAA will probably accept any answer.

36.
0415. What is the approximate base of the cumulus clouds if the temperature at 1,000 feet MSL is 68 °F and the dewpoint is 48 °F?

A— 3,000 feet MSL.
B— 4,000 feet MSL.
C— 6,000 feet MSL.
D— 8,000 feet MSL.

Answer (C) is correct (0415). *(AvW Chap 6)*

The height of cumuliform cloud bases can be estimated using surface temperature/dewpoint spread. Unsaturated air in a convective current cools at about 5.4°F/1,000 ft, and dewpoint decreases about 1°F/1,000 ft. In a convective current, temperature and dewpoint converge at about 4.4°F/1,000 ft. Thus, if the temperature and dewpoint are 68°F and 48°F, respectively, at 1,000 ft, there would be a 20° spread which, divided by the lapse rate of 4.4°/1,000 ft, is approximately 4,500 ft AGL. Adding the 4,500 ft to the 1,000 ft MSL provides the approximate base of the cumulus clouds at 6,000 ft MSL.

Chapter 7: Weather

STABILITY OF AIR MASSES

37.
0412. What is a characteristic of stable air?

A— Stratiform clouds.
B— Unlimited visibility.
C— Fair weather cumulus clouds.
D— Temperature decreases rapidly with altitude.

Answer (A) is correct (0412). *(AvW Chap 6)*
Stratiform clouds, i.e., layer-type clouds, characteristically have stable air.
Answer (B) is incorrect because restricted, not unlimited, visibility near the ground is an indication of stable air. Answer (C) is incorrect because fair weather cumulus clouds indicate unstable conditions. Answer (D) is incorrect because, if temperature decreases rapidly with altitude, the high lapse rate produces lifting activity.

38.
0423. A stable air mass is most likely to have which of these characteristics?

A— Showery precipitation.
B— Turbulent air.
C— Smooth air.
D— Cumuliform clouds.

Answer (C) is correct (0423). *(AvW Chap 6)*
A stable air mass usually has smooth air.
Answers (A), (B), and (D) are incorrect because showery precipitation, turbulent air, and cumuliform clouds are characteristics of an unstable air mass.

39.
0434. Steady precipitation, in contrast to intermittent showers, preceding a front is an indication of

A— cumuliform clouds with moderate turbulence.
B— stratiform clouds with moderate turbulence.
C— cumuliform clouds with little or no turbulence.
D— stratiform clouds with little or no turbulence.

Answer (D) is correct (0434). *(AvW Chap 8)*
Steady precipitation preceding a front is usually an indication of a warm front, which results from warm air being cooled from the bottom by colder air. This results in stratiform clouds with little or no turbulence.
Answers (A) and (C) are incorrect because cumuliform clouds have intermittent rather than steady precipitation. Answer (B) is incorrect because stratiform clouds usually are not turbulent.

40.
0413. Moist, stable air flowing upslope can be expected to

A— produce stratus type clouds.
B— produce a temperature inversion.
C— cause showers and thunderstorms.
D— develop convective turbulence.

Answer (A) is correct (0413). *(AvW Chap 6)*
Moist, stable air flowing upslope produces stratus type clouds as it cools.
Answer (B) is incorrect because temperature inversions mean that warmer air exists aloft, which is unusual, especially around mountain areas with slopes. Answers (C) and (D) are incorrect because thunderstorms and convective turbulence are dependent on unstable, not stable, air.

41.
0422. What are characteristics of unstable air?

A— Turbulence and good surface visibility.
B— Turbulence and poor surface visibility.
C— Nimbostratus clouds and good surface visibility.
D— Nimbostratus clouds and poor surface visibility.

Answer (A) is correct (0422). *(AvW Chap 6)*
The characteristics of unstable air are turbulence and good surface visibility.
Answer (B) is incorrect because poor surface visibility is a characteristic of stable air. Answers (C) and (D) are incorrect because stratus clouds are characteristic of stable air.

42.
0431. An unstable air mass is most likely to have which of these characteristics?

A— Stratiform clouds and fog.
B— Turbulent air.
C— Continuous precipitation.
D— Fair to poor visibility in haze and smoke.

Answer (B) is correct (0431). *(AvW Chap 6)*

An unstable air mass is most likely to have turbulent air. This is in contrast to stable air masses, which likely have stratiform clouds, fog, continuous precipitation, fair to poor visibility, and haze.

43.
0421. A moist, unstable air mass is characterized by

A— cumuliform clouds and showery precipitation.
B— poor visibility and smooth air.
C— stratiform clouds and continuous precipitation.
D— fog and drizzle.

Answer (A) is correct (0421). *(AvW Chap 8)*

A moist, unstable air mass is characterized by cumuliform clouds and showery precipitation.

Answers (B), (C), and (D) are incorrect because stable air masses are usually characterized by poor visibility, smooth air, stratiform clouds, continuous precipitation (e.g., drizzle), and fog.

44.
0410. Which would decrease the stability of an air mass?

A— Warming from below.
B— Cooling from below.
C— Decrease in water vapor.
D— Sinking of the air mass.

Answer (A) is correct (0410). *(AvW Chap 8)*

When air is warmed from below, it tends to rise, resulting in instability; i.e., vertical movement occurs.

Answer (B) is incorrect because cooling from below keeps the surface or the lower area cool, and the air does not rise. Answer (C) is incorrect because, as water vapor in air decreases, the air tends to sink. Answer (D) is incorrect because sinking air masses are cooling air masses, which do not create turbulence.

45.
0414. If an unstable air mass is forced upward, what type clouds can be expected?

A— Clouds with a temperature inversion.
B— Clouds with little vertical development.
C— Clouds with little associated turbulence.
D— Clouds with considerable vertical development and associated turbulence.

Answer (D) is correct (0414). *(AvW Chap 6)*

When unstable air is lifted, it usually results in considerable vertical development and associated turbulence, i.e., convective activity.

Answers (A) and (B) are incorrect because stable rather than unstable air creates layer-type clouds with little vertical development. Answer (C) is incorrect because layer-type clouds usually have little turbulence unless they are lenticular clouds created by mountain waves or other high-altitude clouds associated with high winds near or in the jet stream.

46.
0409. Which measurements can be used to determine the stability of the atmosphere?

A— Atmospheric pressure.
B— Actual lapse rate.
C— Surface temperature.
D— Wind velocity.

Answer (B) is correct (0409). *(AvW Chap 8)*

The stability of the atmosphere is determined by vertical movements of air. Warm air rises when the air above is cooler. The lapse rate, which is the decrease of temperature with altitude, is therefore a measure of stability.

Answers (A), (C), and (D) are incorrect because, while atmospheric pressure, surface temperature, and wind velocity may have some effect on temperature changes and air movements, it is the actual lapse rate that determines the stability of the atmosphere.

Chapter 7: Weather

TEMPERATURE INVERSIONS

47.
0398. A temperature inversion would most likely result in which weather conditions?

A— Clouds with extensive vertical development above an inversion aloft.
B— Good visibility in the lower levels of the atmosphere and poor visibility above an inversion aloft.
C— An increase in temperature as altitude is increased.
D— A decrease in temperature as altitude is increased.

Answer (C) is correct (0398). *(AvW Chap 1)*
By definition, a temperature inversion is a situation in which the temperature increases as altitude increases. The normal situation is that the temperature decreases as altitude increases.
Answer (A) is incorrect because vertical development does not occur in an inversion situation because the warm air cannot rise when the air above is warmer. Answer (B) is incorrect because the inversion traps dust, smoke, and other nuclei beneath the inversion. Answer (D) is incorrect because it is the normal situation, not an inversion.

48.
0396. What feature is associated with a temperature inversion?

A— A stable layer of air.
B— An unstable layer of air.
C— Chinook winds on mountain slopes.
D— Air mass thunderstorms.

Answer (A) is correct (0396). *(AvW Chap 1)*
A temperature inversion is associated with an increase in temperature with height, a reversal of the normal decrease in temperature with height. Thus, any warm air rises to where it is the same temperature and forms a stable layer of air.
Answer (B) is incorrect because instability occurs when the temperature decreases with an increase in height, and the rising air continues to rise. Answer (C) is incorrect because the Chinook winds are dry winds blowing down the eastern slopes of the Rocky Mountains. Answer (D) is incorrect because air mass thunderstorms result from instability and do not occur when there is a temperature inversion.

49.
0397. The most frequent type of ground or surface-based temperature inversion is that produced by

A— terrestrial radiation on a clear, relatively still night.
B— warm air being lifted rapidly aloft in the vicinity of mountainous terrain.
C— the movement of colder air under warm air, or the movement of warm air over cold air.
D— widespread sinking of air within a thick layer aloft resulting in heating by compression.

Answer (A) is correct (0397). *(AvW Chap 1)*
A surface-based temperature inversion is one in which the surface cools the air near the surface and thus it stays still and does not attempt to rise. The main cause of surface cooling is the cooling by terrestrial radiation on clear nights when the ground temperature cools faster than the overlying air and then cools the air next to it.
Answer (B) is incorrect because warm air being lifted rapidly aloft in the vicinity of mountainous terrain generally does not result in a surface-based inversion. Answer (C) is incorrect because the movement of colder air under warm air, which causes an inversion, is caused by a cold front, not terrestrial radiation. Movement of warm air over cold air describes a warm front. Answer (D) is incorrect because sinking of air within a thick layer aloft resulting in heating by compression is not a surface-based inversion.

50.

0399. Which weather conditions should be expected beneath a low-level temperature inversion layer when the relative humidity is high?

A— Smooth air and poor visibility due to fog, haze, or low clouds.
B— Light wind shear and poor visibility due to haze and light rain.
C— Turbulent air and poor visibility due to fog, low stratus type clouds, and showery precipitation.
D— Updrafts and turbulence due to surface heating, fair visibility, and cumulus clouds developing at the top of the inversion.

Answer (A) is correct (0399). *(AvW Chap 1)*
Beneath temperature inversions, there is usually smooth air because there is little vertical movement due to the inversion. There is also poor visibility due to fog, haze, and low clouds (when there is high relative humidity).
Answer (B) is incorrect because wind shears usually do not occur below a low-level temperature inversion. They occur at or just above the inversion. Answers (C) and (D) are incorrect because turbulent air, shower precipitation, updrafts due to surface heating, and cumulus clouds are not present with low-level temperature inversions.

WEATHER BRIEFINGS

51.

0479. When telephoning a weather briefing facility for preflight weather information, pilots should

A— identify themselves as pilots.
B— tell the number of hours they have flown within the preceding 90 days.
C— state the number of occupants on board and the color of the aircraft.
D— state that they possess current medical certificates.

Answer (A) is correct (0479). *(AWS Chap 1)*
When telephoning for a weather briefing you should identify yourself as a pilot so the person can give you an aviation oriented briefing. Many non-pilots call weather briefing facilities to get the weather for other activities.
Answer (B) is incorrect because this is information asked by insurance companies, not FSS. Answer (C) is incorrect because this information is used on the flight plan, not for weather briefings. Answer (D) is incorrect because this is a requirement to act as a pilot-in-command or required crew member, not for a briefing.

52.

0474. When telephoning a weather briefing facility for preflight weather information, pilots should state

A— the number of hours flown within the preceding 90 days.
B— that they possess current medical certificates.
C— whether they intend to fly VFR only.
D— the color of the aircraft and number of occupants on board.

Answer (C) is correct (0474). *(PHAK Chap V)*
When telephoning for a weather briefing, one should identify oneself as a pilot, the route, destination, type of airplane, and whether one intends to fly VFR or IFR to permit the weather briefer to give you the most complete briefing.
Answers (A), (B), and (D) are incorrect because they are not relevant for a weather briefing.

SURFACE WEATHER REPORTS

53.

0395. Ceiling, as used in weather reports, is defined as the height above the Earth's surface of the

A— lowest reported obscuration and the highest layer of clouds reported as overcast.
B— lowest layer of clouds reported as broken or overcast and not classified as thin.
C— lowest layer of clouds reported as scattered, broken, or thin.
D— highest layer of clouds reported as broken or thin.

Answer (B) is correct (0395). *(PHAK Chap V)*
A ceiling is the height above the Earth's surface of the lowest layer of clouds or obscuration phenomena that are reported as broken, overcast, or obscured.
Answers (A) and (D) are incorrect because a ceiling is the lowest layer, not the highest. Answer (C) is incorrect because layers classified as thin, partial, or scattered are not considered ceilings.

Chapter 7: Weather

```
INK SA 1854 CLR 15 106/77/63/1112G18/000
BOI SA 1854 150 SCT 30 181/62/42/1304/015
LAX SA 1852 7 SCT 250 SCT 6HK 129/60/59/2504/991
MDW RS 1856 −X M7 OVC 11/2R+F 990/63/61/3205/980/RF2 RB12
JFK RS 1853 W5 X 1/2F 180/68/64/1804/006/R04RVR22V30 TWR VSBY 1/4
```

FIGURE 32.—Surface Aviation Weather Reports.

54.
0446. (Refer to figure 32.) What are the current conditions depicted for Chicago Midway Airport (MDW)?

A— Thin overcast, measured ceiling 700 feet, overcast 1,100 feet, visibility 2 miles in rain plus fog.
B— Sky partially obscured, measured ceiling 700 overcast, visibility 1-1/2, heavy rain, fog.
C— Thin overcast measured 700 feet overcast, visibility 1-1/2, heavy rain, fog.
D— Sky partially obscured, measured ceiling 700 overcast, visibility 11, occasionally 2, with rain and heavy fog.

Answer (B) is correct (0446). *(PHAK Chap V)*
MDW is reporting sky partially obscured indicated by the -X; a measured 700 ft overcast ceiling indicated by the M7 OVC; visibility is 1½ in heavy rain, indicated by R+; and fog, indicated by F. RS refers to a "record special" weather report that reflects radically different weather from the prior observation.

55.
0447. (Refer to figure 32.) Which reporting stations have VFR weather?

A— All.
B— All except JFK.
C— All except JFK and MDW.
D— INK only.

Answer (C) is correct (0447). *(PHAK Chap V)*
INK is clear with 15 miles visibility, i.e., VFR. BOI is 15,000 ft scattered with 30 miles visibility, i.e., VFR. LAX is 700 ft scattered, 25,000 ft scattered, with 6 miles visibility in haze and smoke, i.e., VFR. MDW is partially obscured, measured 700 overcast, 1½ miles visibility, which is IFR. JFK is indefinite 500, sky obscured, ½ mile visibility in fog, which is IFR.

56.
0448. (Refer to figure 32.) The wind direction and velocity at JFK is from

A— 018° at 4 to 6 knots.
B— 040° variable 22 to 30 knots.
C— 180° at 4 knots.
D— 180° at 40 knots.

Answer (C) is correct (0448). *(PHAK Chap V)*
The wind direction and velocity at JFK are given by the 4 digits together of 1804, which means the wind is from 180° at 4 kts.

57.
0449. (Refer to figure 32.) What are the wind conditions at Wink, Texas (INK)?

A— Calm.
B— 011°, 12 knots gusting to 18 knots.
C— 110°, 12 knots gusting to 18 knots.
D— 111° gusting to 18 knots.

Answer (C) is correct (0449). *(PHAK Chap V)*
At INK, wind conditions are given by 1112G18, which means wind from 110° at 12 kts gusting to 18 kts.

AREA FORECASTS

58.
0470. To best determine forecast weather conditions between weather reporting stations, the pilot should refer to

A— pilot reports.
B— prognostic charts.
C— weather maps.
D— Area Forecasts.

Answer (D) is correct (0470). *(PHAK Chap V)*

An area forecast is a prediction of general weather conditions over an area consisting of several states or portions of states. It is used to obtain expected en route weather conditions and also to provide an insight to weather conditions that might be expected at airports where weather reports or forecasts are not issued.

Answers (A) and (C) are incorrect because pilot reports and weather maps provide current weather information, not forecasts. Answer (B) is incorrect because prognostic charts deal with significant rather than all weather conditions.

59.
0471. To determine the freezing level and areas of probable icing aloft, the pilot should refer to the

A— Radar Summary Chart.
B— Weather Depiction Chart.
C— Area Forecast.
D— Surface Analysis.

Answer (C) is correct (0471). *(PHAK Chap V)*

Area forecasts have a special section on probable icing and freezing levels. It is one of the last sections in the report.

Answer (A) is incorrect because it only includes precipitation. Answer (B) is incorrect because a Weather Depiction Chart does not include any icing information. Answer (D) is incorrect because it shows conditions at the surface, not aloft.

60.
0475. The section of the Area Forecast entitled SIG CLDS AND WX contains a

A— summary of cloudiness and weather significant to flight operations broken down by states or other geographical areas.
B— summary of forecast sky cover, cloud tops, visibility, and obstructions to vision along specific routes.
C— statement of AIRMET's and SIGMET's still in effect at the time of issue.
D— summary of only those clouds and weather considered adverse to safe flight operations.

Answer (A) is correct (0475). *(PHAK Chap V)*

The SIG CLDS AND WX is the clouds and weather plus categorical outlook section, which contains a summary of cloudiness and weather significant to flight operations broken down by states or other geographical areas.

Answer (B) is incorrect because a summary of forecast sky cover, cloud tops, visibility, and obstructions to vision along specific routes is part of a transcribed weather broadcast. Answer (C) is incorrect because AIRMETS and SIGMETS are listed at the end of the FA (area forecast). Answer (D) is incorrect because the summary of only those clouds and weather considered adverse to safe flight operations is part of the flight precaution statement.

TERMINAL FORECASTS

61.
0451. From which primary source should information be obtained regarding expected weather at an airport?

A— Low Level Prog Chart.
B— Weather Depiction Chart.
C— Terminal Forecast.
D— Radar Summary and Weather Depiction Chart.

Answer (C) is correct (0451). *(PHAK Chap V)*

The Terminal Forecast provides a 24-hour advance forecast in 3 to 6 hour blocks for selected terminals.

Answer (A) is incorrect because Low Level Prog Charts forecast weather conditions expected to exist 12 hours and 24 hours in advance for the entire country. They should be used for en route analysis. Answer (B) is incorrect because a Weather Depiction Chart is a national map giving a quick picture of the weather conditions that exist at times stated on the chart, i.e., not a forecast. Answer (D) is incorrect because Radar Summary and Weather Depiction Charts generally predict precipitation only and, again, are national maps.

Chapter 7: Weather

```
OK FT 011447

GAG FT 011515 100 SCT 250 SCT 2610. 16Z 60 SCT C100 BKN 3315G22 CHC C50 BKN
    5TRW. 01Z 250 SCT 3515G25. 09Z VFR WIND..

HBR FT 011515 C120 BKN 250 BKN 3010. 17Z 100 SCT C250 BKN 3215G25 CHC C30 BKN
    3TRW. 00Z 250 SCT 3515G25. 09Z VFR WIND..

MLC FT 011515 C20 BKN 1815 BKN OCNL SCT. 20Z C30 BKN 1815G22 CHC C20 BKN
    1TRW. 03Z C30 BKN 2015 CHC C7 X 1/2TRW+G40. 09Z MVFR CIG TRW..

OKC FT 011515 C12 BKN 140 BKN 1815G28 LWR BKN V SCT. 18Z C30 BKN 250 BKN
    2315G25 LWR BKN OCNL SCT CHC C7 X 1/2TRW+G40. 21Z CFP 100 SCT C250 BKN
    3315G25 CHC C30 BKN 5TRW-. 02Z 100 SCT 250 SCT 3515G25. 09Z VFR WIND..

PNC FT 011515 C100 BKN 250 BKN 1810. 16Z CFP 20 SCT C100 BKN 3115 SCT V BKN. 00Z
    250 SCT 3515G25. 09Z VFR WIND..

TUL FT 011515 C20 BKN 1915G22. 19Z C30 BKN 1815G25 CHC 3TRW. 23Z CFP C100 BKN
    250 BKN 3215G25 CHC C30 BKN 5TRW. 09Z VFR WIND..
```

FIGURE 33.—Terminal Forecast.

62.
0465. (Refer to figure 33.) According to the Terminal Forecast for Oklahoma City (OKC), the cold front should pass through

A— between 21Z and 02Z the next day.
B— between 18Z and 21Z with heavy thunderstorms.
C— between 1515Z and 18Z.
D— after 09Z the next day.

Answer (B) is correct (0465). *(PHAK Chap V)*
The Terminal Forecast for Oklahoma City begins with OKC FT 011515, which means the forecast begins at 15Z on the first day of the month and ends 15Z the next day. Each sentence is a new forecast for a new time period. The forecast conditions are expected to have occurred by the time given. The third one starts with 21Z and is followed by CFP, which indicates cold front passage by 21Z. TRW means heavy thunderstorms.

63.
0466. (Refer to figure 33.) What wind conditions are expected at Hobart (HBR) at 16Z?

A— Calm.
B— 115° at 15 knots.
C— 300° at 10 knots.
D— 320° at 15 knots gusting to 25 knots.

Answer (C) is correct (0466). *(PHAK Chap V)*
At HBR at 16Z, the wind is expected to be from 300° at 10 kts, as indicated by "3010." This is in the 1500Z to 1700Z forecast.

64.
0467. (Refer to figure 33.) What ceiling is forecast for Gage (GAG) between 16Z and midnight Z?

A— 100 scattered.
B— 1,000 broken.
C— 6,000 scattered.
D— 10,000 broken.

Answer (D) is correct (0467). *(PHAK Chap V)*
At GAG, between 16Z and 01Z, the forecast is for a broken ceiling of 10,000 as indicated by "C100 BKN".

Answer (A) is incorrect because there is a chance (CHC) of broken, not scattered. Answer (B) is incorrect because "100" means 10,000, not 1,000. Answer (C) is incorrect because the 6,000 scattered is not a ceiling.

126   Chapter 7: Weather

Questions 65 and 66 refer to Figure 33, which is presented on page 125.

65.
0468. (Refer to figure 33.) What is the outlook for weather conditions at McAlester (MLC)?

A— Ceilings 2,000-3,000 feet with southerly winds.
B— Ceiling 700 feet, sky obscured, visibility 1/2 mile in the thundershowers.
C— VFR except in the thundershowers, peak wind gusts 40 knots.
D— Marginal VFR due to low ceilings and thundershowers.

Answer (D) is correct (0468). *(PHAK Chap V)*
   The outlook is found at the end of the FT. "MVFR" means marginal VFR due to ceilings and thundershowers (CIG TRW).

66.
0469. (Refer to figure 33.) The wind condition in the Terminal Forecast outlook for Ponca City (PNC) is

A— missing.
B— for velocities of 25 knots or stronger.
C— for a wind shift from south to northwest.
D— for the wind to change from a gusty condition to calm.

Answer (B) is correct (0469). *(PHAK Chap V)*
   At PNC, the outlook indication is "VFR WIND." The WIND means that winds are expected in excess of 25 kts.

WEATHER DEPICTION CHARTS

67.
0426. (Refer to figure 31.) Of what value is the depiction chart to the pilot?

A— To determine general weather conditions on which to base flight planning.
B— For a forecast of cloud coverage, visibilities, and frontal activity.
C— To determine the frontal trends and air mass characteristics.
D— For an overall view of thunderstorm activity and forecast cloud heights.

Answer (A) is correct (0426). *(PHAK Chap V)*
   The weather depiction chart is prepared from surface aviation weather reports giving a quick picture of weather conditions as of the time stated on the chart. Thus, it presents general weather conditions on which to base flight planning.
   Answers (B) and (C) are incorrect because prog charts provide forecasts about visibilities, ceilings, frontal trends, etc. Answer (D) is incorrect because Radar Summary Charts provide an overall view of thunderstorm activity and cloud heights.

68.
0427. (Refer to figure 31.) The marginal weather in southeast New Mexico is due to

A— reported thunderstorms.
B— 600-foot overcast ceilings.
C— rain showers.
D— low visibility.

Answer (D) is correct (0427). *(PHAK Chap V)*
   Visibility is indicated as being in the vicinity of 4 miles (note the 4 on the left edge of the circle in southeast New Mexico). On the lower right of Figure 35, marginal VFR is described as 3-5 miles visibility.
   Answer (A) is incorrect because the "R" symbol with the lower right-hand leg appearing to have an arrow indicates thunderstorms. They are adverse and dangerous, not marginal. Answer (B) is incorrect because the ceiling height is 6,000, not 600, i.e., add two zeroes. Answer (C) is incorrect because rain showers are indicated by a triangle pointing down.

Chapter 7: Weather

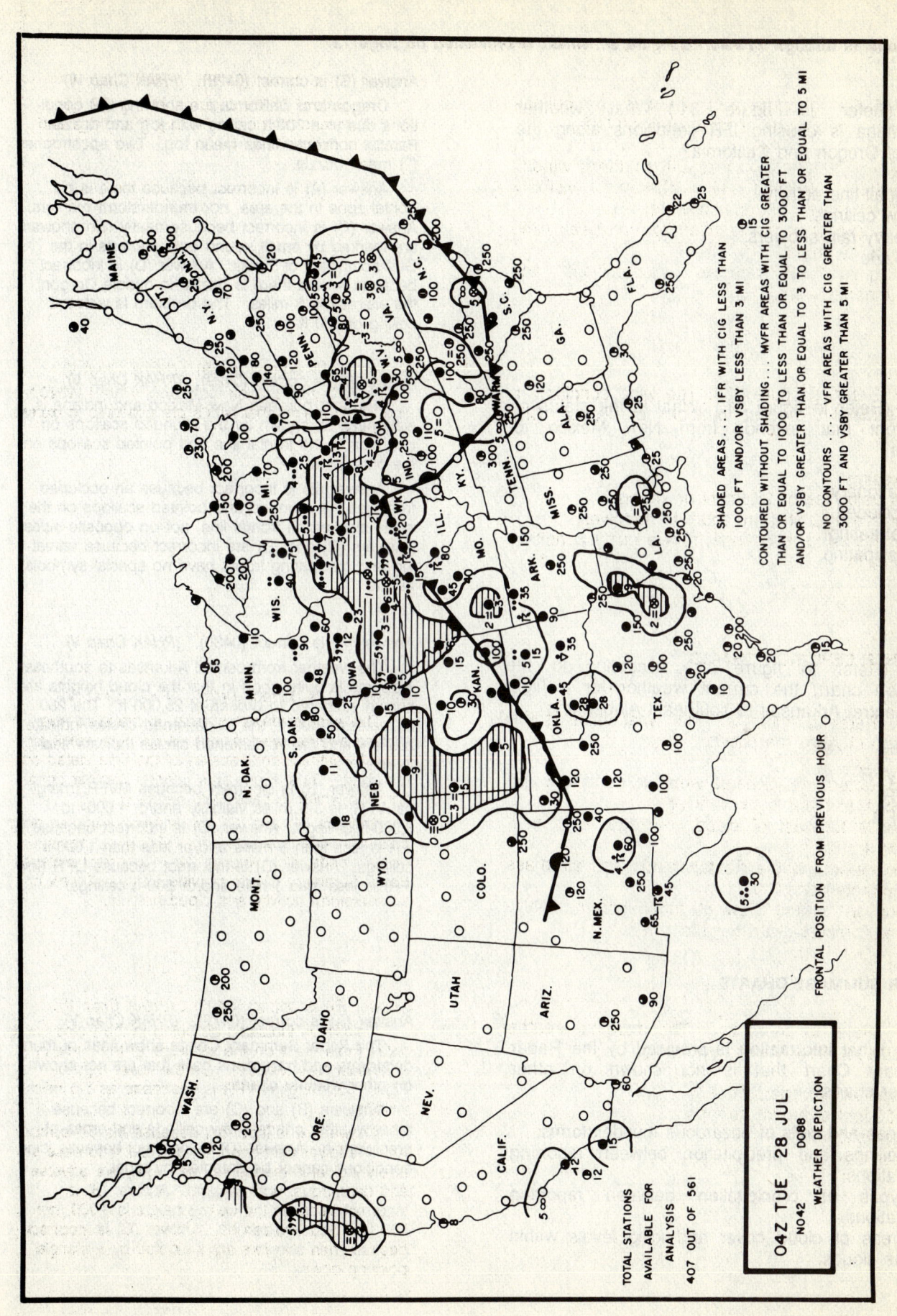

FIGURE 31.—Weather Depiction Chart.

Chapter 7: Weather

Questions 69 through 71 refer to Figure 31, which is presented on page 127.

69.

0428. (Refer to figure 31.) What weather phenomena is causing IFR conditions along the coast of Oregon and California?

A— Squall line activity.
B— Low ceilings.
C— Heavy rain showers.
D— Drizzle.

Answer (B) is correct (0428). *(PHAK Chap V)*

Oregon and California are showing IFR conditions due to a 300-ft ceiling with fog and drizzle. Parallel horizontal lines mean fog. Two apostrophes (") mean drizzle.

Answer (A) is incorrect because there is no frontal zone in the area, nor thunderstorm markers. Answer (C) is incorrect because heavy rain showers are marked by small, round dots such as in the Gulf southwest of Texas. Answer (D) is incorrect because, where there is drizzle in middle Oregon, the visibility is 5 miles. The problem is with the ceiling of 300 ft.

70.

0429. (Refer to figure 31.) What is the status of the front that extends from New Mexico to Indiana?

A— Stationary.
B— Occluded.
C— Retreating.
D— Dissipating.

Answer (A) is correct (0429). *(PHAK Chap V)*

The front between New Mexico and Indiana is stationary, as shown by the rounded scallops on one side of the frontal line and pointed scallops on the other side.

Answer (B) is incorrect because an occluded front has the rounded and pointed scallops on the same side of the frontal line, not on opposite sides. Answers (C) and (D) are incorrect because retreating and dissipating fronts have no special symbols, per se.

71.

0430. (Refer to figure 31.) According to the depiction chart, the current weather for a flight from central Arkansas to southeast Alabama is

A— VFR.
B— MVFR.
C— IFR.
D— LIFR.

Answer (A) is correct (0430). *(PHAK Chap V)*

The weather from central Arkansas to southeast Alabama is quite good in that the cloud heights are shown scattered to broken at 25,000 ft. The 250 indicates 25,000 ft; the ¾ darkened circles indicate broken, and the ¼ darkened circles indicate scattered.

Answer (B) is incorrect because MVFR (marginal VFR) is 3-5 miles visibility and/or 1,000- to 3,000-ft ceilings. Answer (C) is incorrect because IFR is less than 3 miles and/or less than 1,000-ft ceilings. Answer (D) is incorrect because LIFR (low IFR) is less than 1 mile and/or 500-ft ceilings.

RADAR SUMMARY CHARTS

72.

0477. What information is provided by the Radar Summary Chart that is not shown on other weather charts?

A— Lines and cells of hazardous thunderstorms.
B— Ceilings and precipitation between reporting stations.
C— Types of precipitation between reporting stations.
D— Areas of cloud cover and icing levels within the clouds.

Answer (A) is correct (0477). *(PHAK Chap V)*

The Radar Summary Charts show lines of thunderstorms and hazardous cells that are not shown on other weather charts.

Answers (B) and (C) are incorrect because other weather charts show ceilings and areas of precipitation. Answer (D) is incorrect because icing conditions cannot be detected by radar.

Chapter 7: Weather

73.
0450. Radar weather reports are of special interest to pilots because they report

A— large areas of low ceilings and fog.
B— location of precipitation along with type, intensity, and trend.
C— location of broken to overcast clouds.
D— icing conditions.

Answer (B) is correct (0450). *(PHAK Chap V)*

Radar weather reports are of special interest to pilots because they report intense weather, e.g., thunderstorm activity and movement, as well as other forms of precipitation.

Answers (A), (C), and (D) are incorrect because radar does not detect clouds, fog, icing conditions, etc. Radar only detects precipitation.

EN ROUTE FLIGHT ADVISORY SERVICE

74.
0500. En Route Flight Advisory Service (EFAS) is a service specifically designed to provide en route aircraft with

A— filing or closing of flight plans.
B— position reporting.
C— complete preflight briefing.
D— timely and meaningful weather advisories pertinent to the type of flight intended, route of flight and altitude.

Answer (D) is correct (0500). *(AIM Para 503)*

En Route Flight Advisory Service provides en route aircraft with timely and meaningful weather advisories pertinent to the type of flight intended, route of flight, and altitude. It is also a collection and distribution point for pilot-reported weather information. Information is available in most of the United States for altitudes above 5,000 ft.

Answers (A) and (C) are incorrect because EFAS is not intended to provide general FSS services. These should be conducted on a frequency other than 122.0. Answer (B) is incorrect because position reporting is usually done to ATC, rather than FSS.

75.
0499. Contact with an En Route Flight Advisory Service Station (EFAS) may be established by calling

A— EFAS on 122.2 MHz.
B— Flight Assistance on 122.5 MHz.
C— Flight Watch on 122.0 MHz.
D— AIRMET on any Flight Watch frequency.

Answer (C) is correct (0499). *(AIM Para 503)*

En Route Flight Advisory Service is contacted by calling "Flight Watch" on 122.0 and indicating your approximate position.

Answer (A) is incorrect because 122.2 is a common en route FSS frequency. Answer (B) is incorrect because there is no concept of flight assistance, although 122.5 is a flight service frequency. Answer (D) is incorrect because AIRMET refers to "Airmen's Meteorological Information." It is an in-flight weather advisory to amend area forecasts concerning weather phenomena that are of operational interest to all aircraft and potentially hazardous to aircraft having limited capability because of lack of equipment, instrumentation, or pilot qualifications. AIRMETs concern weather of less severity than that covered by SIGMETs.

130 Chapter 7: Weather

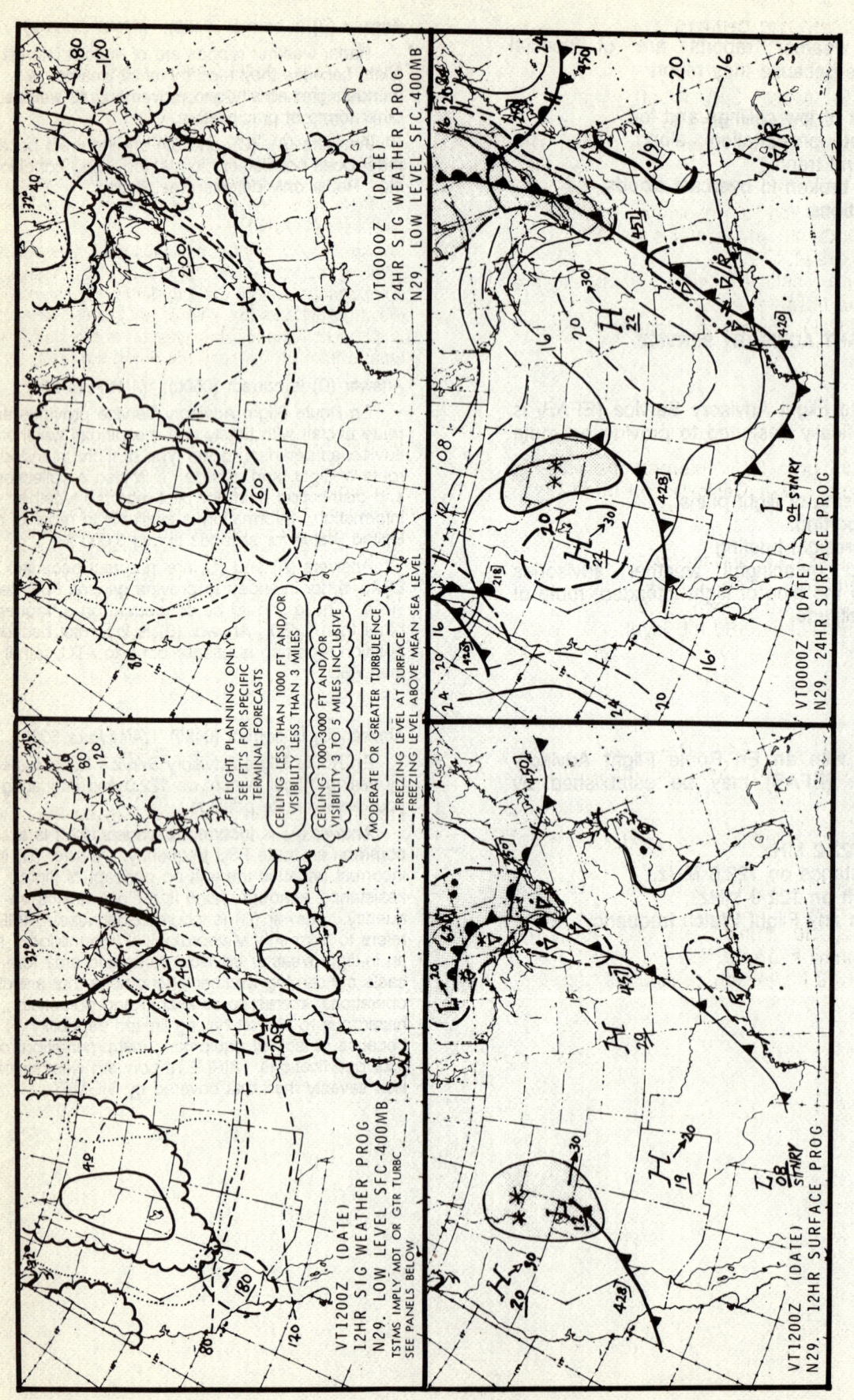

FIGURE 34.—Significant Weather Prognostic Chart.

Chapter 7: Weather

LOW-LEVEL PROGNOSTIC CHARTS

76.

0480. (Refer to figure 34.) What weather is forecast for the Gulf Coast area just ahead of the cold front during the first 12 hours?

A— Marginal VFR to IFR with intermittent thundershowers and rain showers.
B— IFR with moderate or greater turbulence over the coastal areas.
C— Thunderstorm cells moving northeastward ahead of the front.
D— Rain and drizzle dissipating, clearing along the front.

Answer (A) is correct (0480). *(PHAK Chap V)*

During the first 12 hours in the Gulf Coast area just ahead of the cold front which extends from the southern tip of Illinois down through southern Texas (bottom panel), the weather is forecast as marginal VFR to IFR (top panel). Also, there is an area of thunderstorms over southern Louisiana and Mississippi (bottom panel).

Answer (B) is incorrect because the turbulence is along the frontal line not ahead of it. Answer (C) is incorrect because the thunderstorms are not moving northeast but east or southeast. Answer (D) is incorrect because the question is directed to the area in front of the front, not along the front.

77.

0481. (Refer to figure 34.) Interpret the weather symbol depicted in lower California on the 12-hour Significant Weather Prog.

A— Moderate turbulence, surface to 18,000 feet.
B— Thunderstorm tops at 18,000 feet.
C— Base of clear air turbulence, 18,000 feet.
D— Moderate turbulence, 180 mb level.

Answer (A) is correct (0481). *(PHAK Chap V)*

On the 12-hour Significant Weather Prog in southern California, they are forecasting moderate turbulence as designated by the symbol of a small peaked hat. Note the broken line indicates moderate or greater turbulence. The peaked hat identifies it as moderate turbulence. It extends up to 18,000 ft as designated by the line under 180. This is not the base of the clear air turbulence because a line over a number indicates a base.

78.

0482. (Refer to figure 34.) The band of IFR weather associated with the cold front in the western states is forecast to move

A— southeast at 30 knots with moderate snow showers.
B— northeast at 12 knots with the front and producing snow showers.
C— eastward at 30 knots with the low and producing snow showers.
D— eastward at 30 knots with continuous snow.

Answer (D) is correct (0482). *(PHAK Chap V)*

In the lower left panel, the low is going to move to the east at 30 kts as indicated by the arrow and the 30. There are going to be continuous snow showers as indicated by the double asterisk. Two symbols indicate continuous whereas one indicates intermittent.

132 Chapter 7: Weather

This is a Transcribed Weather Broadcast from Oklahoma City, Oklahoma, with Oklahoma City to Amarillo, Texas route. Valid until 23Z.

"Synopsis: Rapidly developing upper level low-pressure system over eastern New Mexico moving eastward over the area through the period. A strong low-level easterly flow south of an artic high-pressure center in Nebraska continuing over the area."

'Adverse conditions: IFR conditions due to low ceilings and visibilities for the west portion of the route spreading east over the entire route by midafternoon. Occasional moderate to severe icing in the clouds and precipitation from the surface to 8 thousand MSL for the west portion of the route. Occasional moderate turbulence 20 to 30 thousand along the route due to the jet stream. Low-level wind shear along the route due to strong surface winds.'

'Route forecast, Oklahoma City–Amarillo: The west portion of the route, ceilings at or below 1 thousand and visibilities occasionally below 3 miles in snow, ice pellets, and fog. The east portion of the route, ceilings 3 thousand becoming ceilings at or below 1 thousand, and visibilities occasionally below 3 miles in snow and fog by midafternoon. Tops to 17 thousand MSL, also scattered to broken clouds at or above 25 thousand MSL along the entire route.'

'Winds aloft at Oklahoma City: 3 thousand 050 at 30, 6 thousand 070 at 50, 9 thousand 110 at 40, 12 thousand 230 at 40, 18 thousand 230 at 65, 24 thousand 240 at 75, 30 thousand 240 at 87, 34 thousand 240 at 80, 39 thousand 240 at 60. Amarillo: 3 thousand 080 at 40, 6 thousand 090 at 45, 9 thousand 120 at 35, 12 thousand 230 at 40, 18 thousand 240 at 70, 24 thousand 240 at 85, 30 thousand 250 at 96, 34 thousand 250 at 90, 39 thousand 240 at 60.'

'Radar reports: At 1435Z, Amarillo radar shows an area of light snow and ice pellets with 7 tenths coverage extending from Amarillo to Gage to Childress to Amarillo. The area has increased in size from past hour. Maximum tops are uniform at 14 thousand MSL. The area is moving to the northeast at 35 knots.'

'Surface weather reports: Amarillo, 1520Z, measured ceiling 8 hundred overcast, visibility 3 miles in light snow and fog, temperature 25, dew point 23, wind 090 at 25 gusting to 37, altimeter 29 point 67, the ice pellets began at 25 after the hour and ended 40 minutes past the hour; Gage, 1515Z, measured ceiling 3 hundred broken, 10 thousand overcast, visibility 2 miles in light snow and fog, temperature 22, dew point 20, wind 070 at 18 gusting to 30, altimeter 29 point 97. Childress, 5 hundred scattered, measured ceiling 9 hundred overcast, visibility 6 miles in light snow/ice pellets/fog, temperature 30, dew point 28, wind 110 at 25 gusting to 35, altimeter 29 point 88. Hobart, measured ceiling 25 hundred broken, 5 thousand overcast, visibility 7 miles, temperature 27, dew point 24, wind 080 at 20 gusting to 32, altimeter 29 point 94, snow showers of unknown intensity to the west. Oklahoma City, measured ceiling 3 thousand broken, 8 thousand overcast, visibility 7 miles, temperature 21, dew point 16, wind 040 at 15 gusting to 28, altimeter 30 point 02.'

'Pilot reports: 30 miles east of Amarillo, a pilot of a Boeing 727 during a climbout at Amarillo reported moderate rime icing from the surface to 7 thousand, tops of clouds 15 thousand with cirrus above and moderate turbulence from the surface to 5 thousand."

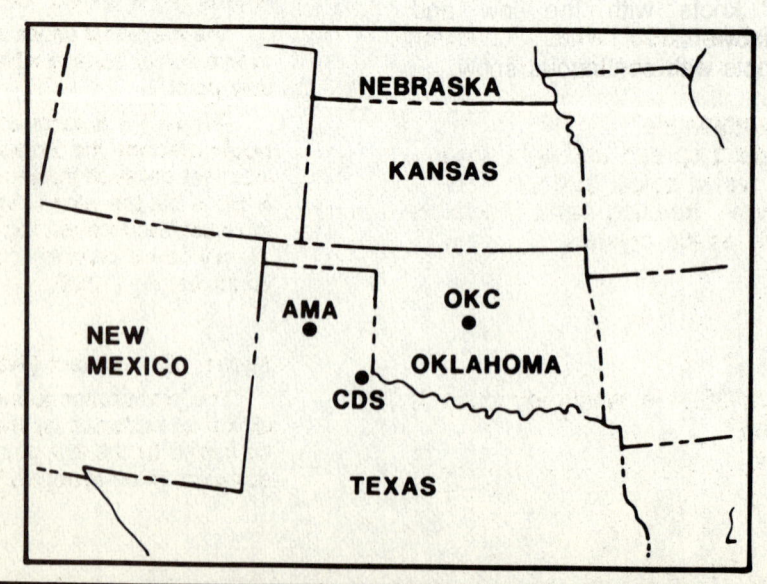

FIGURE 35.—Weather Reference Chart.

Chapter 7: Weather

TRANSCRIBED WEATHER BROADCASTS (TWEBs)

79.
0483. (Refer to figure 35.) What weather phenomenon reported in the TWEB indicates a layer of above freezing temperatures above 7 or 8 thousand?

A— Snow.
B— Ice pellets.
C— Wind shift aloft.
D— Temperature/dewpoint spread.

Answer (B) is correct (0483). *(AvW Chap 5)*

The temperatures at the surface are below freezing. In order to have ice pellets, higher than freezing temperatures are necessary. Thus, ice pellets are caused by rain falling into colder air.

Answer (A) is incorrect because snow is usually formed in less than freezing temperatures. Answer (C) is incorrect because the wind shift aloft does not guarantee that there are warmer temperatures above as the ice pellets do. Answer (D) is incorrect because the temperature/dewpoint spread is at the surface and provides no information about temperatures aloft.

80.
0484. (Refer to figure 35.) The boundary between the air from the high in Nebraska and the low in New Mexico as described by the TWEB is between

A— the surface and 6,000 feet.
B— 6,000 feet and 9,000 feet.
C— 9,000 feet and 12,000 feet.
D— 12,000 feet and 15,000 feet.

Answer (C) is correct (0484). *(AvW Chap 8)*

The boundary between the air from the high in Nebraska and low in New Mexico is between 9,000 and 12,000 ft as indicated by a wind shift from 110° to 230° at Oklahoma City and 120° to 230° at Amarillo between 9,000 and 12,000 ft in the winds aloft. The wind shift indicates different pressure systems.

81.
0485. (Refer to figure 35.) What type clouds normally produce the icing mentioned by the PIREP in the TWEB?

A— Cumulus.
B— Stratocumulus.
C— Stratus.
D— Cumulonimbus.

Answer (C) is correct (0485). *(AvW Chap 10)*

Rime icing generally occurs in stratus clouds whereas clear icing occurs in stratocumulus and cumulonimbus clouds.

82.
0486. (Refer to figure 35.) What condition described in the TWEB is causing the fog?

A— Precipitation and turbulence.
B— Temperature/dewpoint spread and light wind.
C— Moist air moving over a colder surface.
D— Temperature below freezing and moisture sublimating directly as ice crystals.

Answer (C) is correct (0486). *(AvW Chap 12)*

The traditional cause of fog is moist air moving over a colder surface where the air is cooled to its dew point.

Answer (A) is incorrect because the turbulence would preclude the formation of fog. Answer (B) is incorrect because the temperature/dewpoint spread is close but the wind is very strong. Answer (D) is incorrect because ice fog (moisture sublimating directly as ice crystals) occurs in much colder temperatures, e.g., -25°F.

83.
0487. (Refer to figure 35.) In which direction is the precipitation moving?

A— North.
B— Northeast.
C— East.
D— Southeast.

Answer (B) is correct (0487). *(AvW Chap 12)*

The precipitation is moving to the northeast at 35 kts as explained by the radar reports. (Do not be misled by the low pressure system moving eastward in the synopsis.)

WINDS AND TEMPERATURES ALOFT FORECASTS

```
FD WBC 151745
BASED ON 151200Z DATA
VALID 1600Z FOR USE 1800-0300Z. TEMPS NEG ABV 24000

FT     3000    6000     9000     12000    18000    24000    30000    34000    39000
ALS            2420     2635-08  2535-18  2444-30  245945   246755   246862
AMA            2714     2725+00  2625-04  2531-15  2542-27  265842   256352   256762
DEN                     2321-04  2532-08  2434-19  2441-31  235347   236056   236262
HLC            1707-01  2113-03  2219-07  2330-17  2435-30  244145   244854   245561
MKC    0507    2006+03  2215-01  2322-06  2338-17  2348-29  236143   237252   238160
STL    2113    2325+07  2332+02  2339-04  2356-16  2373-27  239440   730649   731960
```

FIGURE 30.—Winds Aloft Forecast.

84.

0424. (Refer to figure 30.) What wind is forecast for St. Louis (STL) at 6,000 feet?

A— 210° at 13 knots.
B— 230° at 25 knots.
C— 230° at 25 knots, peak gust 7 minutes past the hour.
D— 232° at 5 knots.

Answer (B) is correct (0424). *(PHAK Chap V)*

The wind forecast at STL at 6,000 ft is 2325 + Ø7, which is 230° at 25 kts and a temperature of +7°C.

85.

0425. (Refer to figure 30.) Interpret the Winds Aloft Forecast for 3,000 feet at Kansas City (MKC).

A— 005° at 7 knots.
B— 050° at 7 knots.
C— 360° at 5 knots.
D— 360° at 50 knots.

Answer (B) is correct (0425). *(PHAK Chap V)*

The wind forecast for Kansas City (MKC) at 3,000 ft is 0507, which is 050° at 7 kts. The temperature is "missing" because no temperatures are forecast for the 3,000-ft level or for levels within 2,500 ft AGL.

86.

0478. What values are used for Winds Aloft Forecasts?

A— Magnetic direction and knots.
B— Magnetic direction and miles per hour.
C— True direction and knots.
D— True direction and miles per hour.

Answer (C) is correct (0478). *(PHAK Chap V)*

For Winds Aloft Forecasts, the wind direction is given in true direction and the windspeed is in kts.

Chapter 7: Weather

AIRMETS AND SIGMETS

87.
0472. SIGMET's are issued as a warning of weather conditions hazardous

A— particularly to light aircraft.
B— to all aircraft.
C— only to light aircraft operations.
D— particularly to heavy aircraft.

Answer (B) is correct (0472). *(PHAK Chap V)*
SIGMETs warn of weather considered potentially hazardous to all categories of aircraft. SIGMETs are forecasts of tornadoes, lines of thunderstorms, embedded thunderstorms, large hail, severe and extreme turbulence, severe icing, and widespread sandstorms and snowstorms.

88.
0473. AIRMET's are issued as a warning of weather conditions hazardous

A— to all airplanes.
B— particularly to light airplanes.
C— to VFR operations only.
D— particularly to heavy airplanes.

Answer (B) is correct (0473). *(PHAK Chap V)*
AIRMETs are issued to advise pilots of weather which may be hazardous to light aircraft (and in some cases all aircraft). Thus, it applies particularly to light airplanes, not all airplanes, nor to VFR operations only, and not particularly to heavy airplanes.

89.
0476. What information is contained in a convective SIGMET in the conterminous United States?

A— Tornadoes, embedded thunderstorms, and hail 3/4 inch or greater in diameter.
B— Severe icing, severe turbulence, or widespread dust storms lowering visibility to less than 3 miles.
C— Weather less than basic VFR and sustained winds of 30 knots or greater at the surface.
D— Ceilings less than 500 feet and visibility less than 1 mile.

Answer (A) is correct (0476). *(AIM Glossary)*
Convective SIGMETs are issued for tornadoes, lines of thunderstorms, embedded thunderstorms, and large hail.
Answer (B) is incorrect because such weather phenomena are covered by SIGMETs. Answer (C) is incorrect because sustained winds of 30 kts or more at the surface and less than basic VFR would be covered in an AIRMET. Answer (D) is incorrect because it describes LIFR, which is not reported in SIGMETs.

Blank Page

CHAPTER EIGHT
FEDERAL AVIATION REGULATIONS

21.181	Duration of Airworthiness Certificates	(2 questions)	138, 147
61.3	Requirements for Certificates, Rating, and Authorizations	(3 questions)	138, 147
61.23	Duration of Medical Certificates	(1 question)	138, 148
61.56	Flight Review	(3 questions)	138, 148
61.57	Recent Flight Experience	(3 questions)	139, 149
61.60	Change of Address	(1 question)	139, 150
61.101	Recreational Pilot Privileges and Limitations	(18 questions)	139, 150
71.5	Extent of Federal Airways	(1 question)	140, 155
71.11	Control Zones	(1 question)	140, 155
91.3	Responsibility and Authority of Pilot-in-Command	(2 questions)	140, 156
91.5	Preflight Action	(4 questions)	140, 156
91.11	Alcohol or Drugs	(1 question)	141, 157
91.14	Use of Safety Belts	(1 question)	141, 157
91.15	Parachutes and Parachuting	(3 questions)	141, 158
91.22	Fuel Requirements for Flight Under VFR	(1 question)	141, 158
91.27	Civil Aircraft Certifications Required	(2 questions)	141, 159
91.29	Civil Aircraft Airworthiness	(1 question)	141, 159
91.31	Civil Aircraft Flight Manual, Marking, and Placard Requirements	(1 question)	141, 159
91.39	Restricted Category Civil Aircraft; Operating Limitations	(1 question)	142, 160
91.52	Emergency Locator Transmitters	(1 question)	142, 160
91.67	Right-of-Way Rules	(6 questions)	142, 160
91.70	Aircraft Speed	(2 questions)	143, 162
91.71	Acrobatic Flight	(4 questions)	143, 162
91.79	Minimum Safe Altitudes	(4 questions)	143, 163
91.90	Terminal Control Areas	(1 question)	143, 164
91.95	Restricted and Prohibited Areas	(1 question)	144, 164
91.105	Basic VFR Weather Minimums	(12 questions)	144, 165
91.109	VFR Cruising Altitude or Flight Level	(3 questions)	145, 167
91.163	Maintenance, Preventive Maintenance, and Alterations	(1 question)	145, 168
91.167	Operation After Maintenance, Preventive Maintenance, Rebuilding, or Alteration	(2 questions)	145, 168
91.169	Inspections	(1 question)	146, 168
91.173	Maintenance Records	(3 questions)	146, 169
NTSB 830	National Transportation Safety Board (Part 830)	(7 questions)	146, 169

138   Chapter 8: Federal Aviation Regulations

This chapter contains outlines of major concepts tested, all FAA test questions and answers regarding Federal Aviation Regulations (FARs), and an explanation of each answer. The subtopics or modules within this chapter are listed above, followed in parentheses by the number of questions from the FAA written test pertaining to that particular module. The two numbers following the parentheses are the page numbers on which the outline and questions begin for that module.

CAUTION: Recall that the sole purpose of this book is to expedite your passing the FAA written test for the recreational pilot certificate. Accordingly, all extraneous material (i.e., topics or regulations not directly tested on the FAA written test) is omitted, even though much more information and knowledge are necessary to fly safely. This additional material is presented in *RECREATIONAL PILOT FLIGHT MANEUVERS* and *PRIVATE PILOT HANDBOOK*, available from Aviation Publications, Inc. See pages 208 to 210 for more information and an order form.

21.181 DURATION OF AIRWORTHINESS CERTIFICATES (2 questions)

1. Airworthiness certificates remain in force as long as maintenance and alteration of the aircraft are performed per FARs.

61.3 REQUIREMENTS FOR CERTIFICATES, RATING, AND AUTHORIZATIONS (3 questions)

1. When acting as a pilot-in-command or as a required pilot flight crewmember, you must have current and appropriate pilot and medical certificates in your personal possession.

61.23 DURATION OF MEDICAL CERTIFICATES (1 question)

1. Class III Medical Certificates (for private and student pilots) expire on the last day of the 24th month after the month of the examination date.

61.56 FLIGHT REVIEW (3 questions)

1. Flight review encompasses a review of Part 91, general operating and flight rules, and those maneuvers necessary, in the opinion of the CFI, to demonstrate safe exercise of the pilot certificate.
2. **Recreational pilots** and **non-instrument-rated private pilots** with less than 400 hrs of flight time must have annual flight reviews consisting of at least 1 hr of ground instruction and 1 hr of flight instruction.
    a. Other pilots must have completed a biennial flight review since the beginning of the preceding 24th calendar month and have a logbook endorsement stating satisfactory completion of the flight review.
    b. Flight reviews must be completed prior to the end of the same month that you received your pilot certificate or previously completed a flight review.
3. A new rating or certificate based on a practical or simulator flight test takes the place of a flight review.

Chapter 8: Federal Aviation Regulations

139

61.57 RECENT FLIGHT EXPERIENCE (3 questions)

1. To carry passengers, you must have made three takeoffs and three landings within the preceding 90 days.
    a. All three landings must be made in aircraft of the same category and class as the one in which passengers are to be carried.
        1) The categories are airplane, rotorcraft, glider, and lighter-than-air.
        2) The classes are single-engine land, single-engine sea, multiengine land, and multiengine sea.
    b. The landings must be to a full stop if the airplane is tailwheel (conventional) rather than nosewheel.

61.60 CHANGE OF ADDRESS (1 question)

1. You must notify the FAA Airman Certification Branch in writing of any change in your permanent mailing address (write to PO Box 25082 • Oklahoma City, OK • 73125).
2. You may not exercise the privileges of your pilot certificate (act as pilot-in-command) after 30 days from moving unless you make this notification.

61.101 RECREATIONAL PILOT PRIVILEGES AND LIMITATIONS (18 questions)

1. A recreational pilot may
    a. **Not** carry more than one passenger.
    b. Share the operating expenses of a flight with the passenger.
    c. Act as pilot-in-command within 50 nautical miles (NM) of an airport at which the pilot has received ground and flight instruction.
    d. Land at an airport within 50 NM of the departure airport.
2. Recreational pilots must carry a logbook with an endorsement attesting to the required instructions.
3. A recreational pilot may not act as pilot-in-command of an aircraft:
    a. That is certificated
        1) For more than four occupants.
        2) With more than one powerplant.
        3) With a powerplant of more than 180 horsepower.
        4) With retractable landing gear.
    b. That is classified as a glider, airship, or balloon.
    c. That is carrying a passenger or property for compensation or hire.
    d. For compensation or hire.
    e. In furtherance of a business.
    f. Between sunset and sunrise.
    g. In airspace in which communication with ATC is required.
    h. At an altitude of more than 10,000 ft MSL or 2,000 ft AGL, whichever is higher.
    i. When the flight or surface visibility is less than 3 statute miles (in either controlled or uncontrolled airspace).

140   Chapter 8: Federal Aviation Regulations

      j.   Without visual reference to the surface.
      k.   On a flight outside the United States.
      l.   To demonstrate that aircraft in flight to a prospective buyer.
      m.   That is used in a passenger-carrying airlift and sponsored by a charitable organization.
      n.   That is towing any object.

4. Recreational pilots, however, when training for a pilot certificate, may fly solo under a CFI's supervision and with logbook signoffs.
      a.   In an aircraft, not currently rated.
      b.   At night but not with less than 5 miles visibility.
      c.   In airspace requiring ATC control.

5. A recreational pilot may **not** act as a required pilot flight crewmember on any aircraft for which more than one pilot is required.

6. A recreational pilot who has less than 400 flight hours and who has not logged pilot-in-command time within the preceding 180 days may **not** act as pilot-in-command until receiving flight instruction and logbook endorsement (re: competency).

71.5 EXTENT OF FEDERAL AIRWAYS (1 question)

1. Federal airways include the
      a.   Airspace extending upward from 1,200 ft AGL to, but not including, 18,000 ft MSL; and the
      b.   Airspace within parallel boundary lines 4 miles each side of the airway's center line.

71.11 CONTROL ZONES (1 question)

1. Control zones extend upward from the surface to the base of the continental control area, or 14,500 ft MSL.
2. The continental control area exists over the 48 contiguous United States, the District of Columbia, and most of Alaska.

91.3 RESPONSIBILITY AND AUTHORITY OF PILOT-IN-COMMAND (2 questions)

1. In emergencies, a pilot may deviate from the FARs to the extent needed to maintain the safety of the airplane and passengers.
2. The pilot-in-command of an aircraft is directly responsible for, and is the final authority as to, the operation of that aircraft.
3. A written report of any deviation from FARs should be filed with the FAA upon request.

91.5 PREFLIGHT ACTION (4 questions)

1. Pilots are required to familiarize themselves with all available flight information prior to every flight, and specifically to determine
      a.   Runway lengths at airports of intended use and the airplane's takeoff and landing requirements.

Chapter 8: Federal Aviation Regulations

91.11 ALCOHOL OR DRUGS (1 question)

1. No person may act as a crewmember of a civil airplane while having .04% by weight or more alcohol in the blood or if any alcoholic beverages have been consumed within the preceding 8 hrs.

91.14 USE OF SAFETY BELTS (1 question)

1. Pilots must ensure that each occupant is briefed on how to use the seatbelts.
2. Pilots must notify all occupants to fasten their seatbelts before taking off or landing.
3. All occupants of airplanes must wear seatbelts during takeoffs and landings.

91.15 PARACHUTES AND PARACHUTING (3 questions)

1. Each occupant of an aircraft must wear an approved parachute during acrobatic maneuvers.
    a. Banked turns in excess of 60° are acrobatic.
2. A chair-type parachute must have been packed by a certificated parachute-rigger within the preceding 120 days.

91.22 FUEL REQUIREMENTS FOR FLIGHT UNDER VFR (1 question)

1. During the day, FARs require fuel sufficient to fly to the first point of intended landing and then for an additional 30 minutes, assuming normal cruise speed.
    a. Recreational pilots can only fly at night solo under a CFI's supervision when training to upgrade.

91.27 CIVIL AIRCRAFT CERTIFICATIONS REQUIRED (2 questions)

1. Except as provided in § 91.28, no person may operate a civil aircraft unless it has within it the following:
    a. An appropriate and current airworthiness certificate.
    b. A registration certificate issued to its owner.
    c. Operating limitations.
2. No person may operate a civil aircraft unless the airworthiness certificate required by paragraph 1 of this section or a special flight authorization issued under § 91.28 is displayed at the cabin or cockpit entrance so that it is legible to passengers or crew.

91.29 CIVIL AIRCRAFT AIRWORTHINESS (1 question)

1. The pilot-in-command is responsible for determining that the airplane is airworthy prior to every flight.

91.31 CIVIL AIRCRAFT FLIGHT MANUAL, MARKING, AND PLACARD REQUIREMENTS (1 question)

1. The airworthiness certificate, the FAA registration certificate, and the aircraft flight manual or operating limitations must be aboard.

2. The acronym ARROW can be used as a memory aid. The Federal Communications Commission (FCC), not the FAA, requires the radio station license, and weight and balance is considered an operating limitation.

   Airworthiness certificate
   Registration certificate
   Radio station license (FCC requirement)
   Operating limitations, including
   Weight and balance data

   Also, flotation gear if flight over water is expected.

3. The operating limitations of an airplane may be found in the airplane flight manual, approved flight manual material, markings, and placards, or any combination thereof.

91.39 RESTRICTED CATEGORY CIVIL AIRCRAFT; OPERATING LIMITATIONS (1 question)

1. Restricted category civil aircraft may not normally be operated
   a. Over densely populated areas,
   b. In congested airways, or
   c. Near a busy airport where passenger transport is conducted.

91.52 EMERGENCY LOCATOR TRANSMITTERS (1 question)

1. ELT batteries must be replaced (or recharged, if rechargeable) after 1 cumulative hour of use or after 50% of their useful life expires.

91.67 RIGHT-OF-WAY RULES (6 questions)

1. Aircraft in distress have the right-of-way over all other aircraft.

2. When two aircraft are approaching head on or nearly so, the pilot of each aircraft should turn to his/her right, regardless of category.

3. When two aircraft of different categories are converging, the right-of-way depends upon who has the least maneuverability. Thus, the right-of-way belongs to
   a. Balloons over
   b. Gliders over
   c. Airships over
   d. Airplanes or rotorcraft.

4. When aircraft of the same category are converging at approximately the same altitude, except head on or nearly so, the aircraft to the other's right has the right-of-way.
   a. If an airplane of the same category as yours is approaching from your right side, it has the right-of-way.

5. When two or more aircraft are approaching an airport for the purpose of landing, the aircraft at the lower altitude has the right-of-way.
   a. This rule shall not be abused by cutting in front of or overtaking another aircraft.

Chapter 8: Federal Aviation Regulations

91.70 AIRCRAFT SPEED (2 questions)

1. The speed limit is 250 kts (288 mph) when flying below 10,000 ft MSL.
2. In airport traffic areas, the speed limit is 156 kts (180 mph) for airplanes with reciprocating engines, and 200 kts (230 mph) for jet-engine aircraft (BUT 250 kts if the ATA is within a TCA).
3. Under TCAs and in VFR corridors through TCAs, the speed limit is 200 kts (230 mph).

91.71 ACROBATIC FLIGHT (4 questions)

1. Acrobatic flight includes all intentional maneuvers that
    a. Are not necessary for normal flight and
    b. Involve an abrupt change in the airplane's attitude.
2. Acrobatic flight is prohibited
    a. When visibility is less than 3 statute miles,
    b. When altitude is below 1,500 ft AGL,
    c. Within control zones or federal airways, or
    d. Over any congested area or over an open-air assembly of people.

91.79 MINIMUM SAFE ALTITUDES (4 questions)

1. Over congested areas (cities, towns, settlements, or open-air assemblies), the pilot must maintain an altitude of 1,000 ft above the highest obstacle within a horizontal radius of 2,000 ft of the airplane.
2. The minimum altitude over other than congested areas is 500 ft AGL.
    a. Over open water or sparsely populated areas, the airplane may not be operated closer than 500 ft to any person, vessel, vehicle, or structure.
3. Altitude in all areas must be sufficient to permit an emergency landing without undue hazard to persons or property on the surface if a power unit fails.

91.90 TERMINAL CONTROL AREAS (1 question)

1. Terminal Control Areas (TCAs) are controlled airspaces usually found at larger airports with high volumes of traffic.
2. The airplane must have a
    a. VOR receiver,
    b. Two-way communications radio, and a
    c. Transponder equipped with Mode C.
        1) Mode C permits ATC to obtain an altitude readout on their radar screen.
3. Authorization from air traffic control (ATC) is required, regardless of weather conditions.
4. A private pilot certificate is required to operate in a TCA.
    a. Student pilots and recreational pilots under the supervision of a CFI are permitted to operate in TCAs with a proper logbook endorsement.

91.95 RESTRICTED AND PROHIBITED AREAS (1 question)

1. Restricted areas are a type of special use airspace within which your right to fly is limited.
    a. Restricted areas have unusual and often invisible hazards to aircraft (i.e., balloons, military operations, or other dangerous activities).
    b. Although restricted areas are not always in use during the times posted in the legend of sectional charts, permission to fly in that airspace must first be obtained from the controlling agency.
        1) The controlling agency is listed for each restricted area at the bottom of sectional (VFR) charts.

91.105 BASIC VFR WEATHER MINIMUMS (12 questions)

VFR WEATHER MINIMUMS WITHIN CONTROLLED AIRSPACE

1. When 1,200 ft or less AGL, or more than 1,200 ft AGL but less than 10,000 ft MSL
    a. Visibility: 3 statute miles
    b. Distance from clouds
        1) 500 ft below
        2) 1,000 ft above
        3) 2,000 ft horizontally
    c. Memory aid: Be higher (1,000 ft) above and lower (500 ft) below.
2. When above 1,200 ft AGL and at or above 10,000 ft MSL
    a. Visibility: 5 statute miles
    b. Distance from clouds
        1) 1,000 ft above and below
        2) 1 statute mile horizontally
3. Except with a special VFR clearance, no one may operate an aircraft under VFR, in a control zone, and beneath the cloud ceiling if the ceiling is less than 1,000 ft AGL.

VFR WEATHER MINIMUMS OUTSIDE CONTROLLED AIRSPACE

1. When 1,200 ft or less AGL
    a. Visibility: 1 statute mile *(Note: Recreational pilots require a minimum of 3 miles.)*
    b. Clear of clouds
2. When above 1,200 ft AGL and below 10,000 ft MSL
    a. Visibility: 1 statute mile *(Note: Recreational pilots require a minimum of 3 miles.)*
    b. Distance from clouds
        1) 500 ft below
        2) 1,000 ft above
        3) 2,000 ft horizontally

Chapter 8: Federal Aviation Regulations

145

3. When above 1,200 ft AGL and at or above 10,000 ft MSL
   a. Visibility: 5 statute miles
   b. Distance from clouds
      1) 1,000 ft above and below
      2) 1 statute mile horizontally

4. Memory aid: VFR weather minimums for controlled and uncontrolled airspace are the same except that uncontrolled airspace calls for
   a. 1 statute mile visibility if under 10,000 ft MSL
   b. Clear of clouds at 1,200 ft or less AGL

5. No person may takeoff or land at any airport in a control zone under basic VFR unless ground visibility is 3 statute miles and the ceiling is not less than 1,000 ft.
   a. If ground visibility is not reported, then flight visibility must be 3 statute miles.

Remember, FAR 61.101 requires recreational pilots to have 3 miles visibility during the day.

91.109 VFR CRUISING ALTITUDE OR FLIGHT LEVEL (3 questions)

1. Specified altitudes are required for VFR cruising flight at more than 3,000 ft AGL and below 18,000 ft MSL.
   a. The altitude prescribed is based upon the magnetic course (not magnetic heading).
   b. The altitude is prescribed in feet above mean sea level (MSL).
   c. Use an odd thousand-foot MSL altitude plus 500 ft for magnetic courses of 0° to 179°, e.g., 3,500, 5,500, 7,500 ft.
   d. Use an even thousand-foot MSL altitude plus 500 ft for magnetic courses of 180° to 359°, e.g., 4,500, 6,500, or 8,500 ft.
   e. As a memory aid, the "E" in Even does not indicate east; i.e., on easterly heading of 0° through 179°, use odd rather than even.

91.163 MAINTENANCE, PREVENTIVE MAINTENANCE, AND ALTERATIONS (1 question)

1. The owner or operator of an aircraft is primarily responsible for maintaining that aircraft in an airworthy condition.

91.167 OPERATION AFTER MAINTENANCE, PREVENTIVE MAINTENANCE, REBUILDING, OR ALTERATION (2 questions)

1. When aircraft alterations or repairs change the flight characteristics, the aircraft must be test flown and approved for return to service prior to carrying passengers.
   a. The pilot test flying the aircraft must be at least a private pilot and rated for the type of aircraft being tested.

2. All preventive maintenance must be recorded in the aircraft's logbooks, detailing the work done and by whom, before any one may operate that aircraft.

146   Chapter 8: Federal Aviation Regulations

91.169 INSPECTIONS (1 question)

1. Annual inspections expire on the last day of the 12th calendar month after the previous annual inspection.

2. All aircraft that are used for compensation or hire including flight instruction must be inspected on a 100-hr basis in addition to the annual inspection.

    a. One-hundred-hour inspections are due every 100 hrs from the prior due date, regardless as to when the inspection was actually performed.

91.173 MAINTENANCE RECORDS (3 questions)

1. The completion of the annual inspection and the airplane's return to service should be appropriately documented in the airplane maintenance records.

    a. The documentation should include the current status of airworthiness directives and the method of compliance.

NTSB 830 NATIONAL TRANSPORTATION SAFETY BOARD (Part 830) (7 questions)

1. Even when no injuries occur to occupants, an airplane accident resulting in substantial damage must be reported to the nearest National Transportation Safety Board (NTSB) field office immediately.

    a. A written report is also required within 10 days.

2. The following incidents must also be reported immediately to the NTSB:

    a. Inability of any required crewmember to perform normal flight duties because of in-flight injury or illness,
    b. In-flight fire,
    c. Flight control system malfunction or failure,
    d. An overdue airplane that is believed to be involved in an accident,
    e. An airplane collision in flight, and
    f. Turbine (jet) engine failures.

3. Written reports of the above six incidents are only required if requested by the NTSB.

4. Prior to the time the NTSB or its authorized representative takes custody of aircraft wreckage, mail, or cargo, such wreckage, mail, or cargo may not be disturbed or moved except

    a. To remove persons injured or trapped,
    b. To protect the wreckage from further damage, or
    c. To protect the public from injury.

Chapter 8: Federal Aviation Regulations

147

QUESTIONS AND ANSWER EXPLANATIONS

All of the FAA questions from the written test for the Recreational Pilot certificate relating to FARs and the material outlined previously are reproduced below in the same modules as the previous outlines. To the immediate right of each question are the correct answer and answer explanation. You should cover these answers and answer explanations while responding to the questions. Refer to the general discussion in Chapter 1 on how to take the examination.

Remember that the questions from the FAA's Written Test Book have been reordered by topic, and the topics have been organized into a meaningful sequence. Accordingly, the first line of the answer explanation gives the FAA question number and the citation of the authoritative source for the answer.

21.181 DURATION OF AIRWORTHINESS CERTIFICATES

1.
0387. An Airworthiness Certificate remains valid

A— until ownership is transferred.
B— provided the aircraft has not had major damage.
C— until surrendered, suspended, or revoked.
D— provided the aircraft is maintained and operated according to FAR's.

Answer (D) is correct (0387). *(FAR 21.181)*

The airworthiness certificate of an airplane remains valid as long as the airplane is operated and maintained as required by the FARs.

Answer (A) is incorrect because the registration certificate (not the airworthiness certificate) becomes invalid once ownership is transferred. Answer (B) is incorrect because damage is a specific example; the broader coverage of answer (D) is more inclusive. Answer (C) is incorrect because the pilot in command is responsible for determining whether the airplane is in condition for safe flight.

2.
0393. How long does the Airworthiness Certificate of an aircraft remain valid?

A— As long as the aircraft has a current Registration Certificate.
B— Indefinitely, unless the aircraft suffers major damage.
C— As long as the aircraft is maintained and operated as required by FAR's.
D— Indefinitely, unless the prescribed operating limitations are exceeded.

Answer (C) is correct (0393). *(FAR 21.181)*

The airworthiness certificate of an airplane remains valid as long as the airplane is in an airworthy condition, i.e., operated and maintained as required by the FARs.

Answer (A) is incorrect because the registration certificate is the document evidencing ownership. A changed registration has no effect on the airworthiness certificate. Answers (B) and (D) are incorrect because the airplane must be maintained and operated according to the FARs. While damage and operations exceeding prescribed limits may constitute specific infractions of the FARs, answer (C) is more inclusive and therefore the better answer.

61.3 REQUIREMENTS FOR CERTIFICATES, RATING, AND AUTHORIZATIONS

3.
0338. When must a current pilot certificate be in the pilot's personal possession?

A— Anytime when on board an aircraft.
B— When acting as a crew chief during launch and recovery.
C— When a passenger is carried.
D— Anytime when acting as pilot in command or a required crewmember.

Answer (D) is correct (0338). *(FAR 61.3)*

Current and appropriate pilot and medical certificates must be in your personal possession when you act as pilot in command or as a required pilot flight crewmember.

Answer (A) is incorrect because it must be in the pilot's possession anytime flying as PIC or required crewmember. Answer (B) is incorrect because it must be in possession whenever acting as a PIC or required crewmember. Answer (C) is incorrect because anytime you fly as PIC or required crewmember, you must have the certificate in possession regardless if passengers are carried or not.

Chapter 8: Federal Aviation Regulations

4.
0316. A recreational pilot acting as pilot in command, or in any other capacity as a required pilot flight crewmember, must have in his/her personal possession while aboard the aircraft

A— a current logbook endorsement to show that a flight review has been satisfactorily accomplished.
B— the current and appropriate pilot and medical certificates.
C— a current endorsement on the pilot certificate to show that a flight review has been satisfactorily accomplished.
D— the pilot logbook to show recent experience requirements to serve as pilot in command have been met.

Answer (B) is correct (0316). *(FAR 61.3)*
 Current and appropriate pilot and medical certificates must be in your personal possession when you act as pilot in command or as a required pilot flight crewmember.
 Answers (A) and (D) are incorrect because you need not have pilot logbooks in your possession or aboard the airplane, except to show you are within 50 NM of where you received flight instruction. Answer (C) is incorrect because the endorsement after a flight review is made in your pilot logbook, not on your certificate.

5.
0319. What document(s) must be in your personal possession while operating as pilot in command of an aircraft?

A— A certificate showing accomplishment of a current flight review.
B— A certificate showing accomplishment of a checkout in the aircraft and a current flight review.
C— A pilot logbook with endorsements showing accomplishment of a current flight review and recency of experience.
D— An appropriate pilot certificate and a current medical certificate.

Answer (D) is correct (0319). *(FAR 61.3)*
 Current and appropriate pilot and medical certificates must be in your personal possession when you act as pilot in command or as a required pilot flight crewmember.
 Answers (A) and (B) are incorrect because flight reviews and checkouts in aircraft are documented in the pilot logbook rather than on separate certificates. Answer (C) is incorrect because pilots need not have their logbooks in their possession.

61.23 DURATION OF MEDICAL CERTIFICATES

6.
0317. A Third-Class Medical Certificate was issued on August 10, this year. To exercise the privileges of a Recreational Pilot Certificate, the medical certificate will expire on

A— August 10, 2 years later.
B— August 10, 3 years later.
C— August 31, 2 years later.
D— August 31, 3 years later.

Answer (C) is correct (0317). *(FAR 61.23)*
 The Class III or Third-Class medical certificate is good for 2 calendar years. It expires the last day of the month 2 years after it is issued.
 Answer (A) is incorrect because medical certificates expire at the end of the month. Answers (B) and (D) are incorrect because no pilot medical certificate remains in force for 3 years.

61.56 FLIGHT REVIEW

7.
0302. Each recreational pilot who has logged fewer than 400 hours of flight time is required to have

A— a biennial flight review.
B— an annual flight review.
C— a semiannual flight review.
D— a quarterly flight review.

Answer (B) is correct (0302). *(FAR 61.56)*
 Each recreational pilot who has logged fewer than 400 hrs of flight time and each non-instrument rated private pilot who has logged fewer than 400 hrs of flight time must undertake an annual flight review, which will consist of a minimum of 1 hr flight instruction and 1 hr of ground instruction.
 Answer (A) is incorrect because biennial flight reviews are required of recreational pilots who have over 400 or more hours of flight time. Answers (C) and (D) are incorrect because semi-annual and quarterly flight reviews are not required by the FARs.

Chapter 8: Federal Aviation Regulations

8.
0303. If a recreational pilot with less than 400 hours of flight time had a flight review on August 8, this year, when is the next flight review required?

A— August 8, next year.
B— August 8, 2 years later.
C— August 31, next year.
D— August 31, 2 years later.

Answer (C) is correct (0303). *(FAR 61.56)*
FAR 61.56 reads that recreational pilots and non-instrument-rated private pilots who have fewer than 400 hrs of flight time must have completed an annual flight review since the beginning of the 12th month before the month in which the pilot acts as "pilot-in-command." This translates into the requirement of an annual flight review before the end of the same month the next year, e.g., if a flight review is previously undertaken on July 1, it will expire on July 31 one year later.

Answers (A) and (B) are incorrect because both annual and biennial flight reviews expire on the last day of the month, not on the day that the flight review was previously completed. Answers (B) and (D) are incorrect because recreational pilots with less than 400 hrs are required to have an annual, not a biennial, flight review.

9.
0304. If a recreational pilot with more than 400 hours of flight time had a flight review on August 8, this year, when is the next flight review required?

A— August 8, next year.
B— August 8, 2 years later.
C— August 31, next year.
D— August 31, 2 years later.

Answer (D) is correct (0304). *(FAR 61.56)*
FAR 61.56 reads that recreational pilots and non-instrument-rated private pilots who have more than 400 hrs of flight time must have completed a biennial flight review since the beginning of the 24th month before the month in which the pilot acts as "pilot-in-command." This translates into the requirement of a biennial flight review before the end of the same month two years later, e.g., if a flight review is previously undertaken on July 1, it will expire on July 31 two years later.

Answers (A) and (B) are incorrect because both annual and biennial flight reviews expire on the last day of the month, not on the day that the flight review was previously completed. Answers (A) and (C) are incorrect because recreational pilots with more than 400 hrs are required to have a biennial, not an annual, flight review.

61.57 RECENT FLIGHT EXPERIENCE

10.
0321. In general, what experience requirements are needed to carry a passenger?

A— Checkout by a flight instructor.
B— Three takeoffs and landings.
C— Proficiency check by a pilot examiner or an FAA inspector.
D— Application for a waiver.

Answer (B) is correct (0321). *(FAR 61.57)*
In order to carry passengers, pilots must have made, within the preceding 90 days, three takeoffs and landings as the sole manipulator of the controls in the same category and class of aircraft. If a type rating is required, it must be in the same type of aircraft.

Answer (A) is incorrect because checkout by a flight instructor is required for high-performance aircraft, not to meet currency requirements to carry passengers. Answer (C) is incorrect because a proficiency check by a pilot examiner/FAA inspector is required for a new rating and/or certificate. Answer (D) is incorrect because waivers are issued for exceptions to FARs, not pilot currency.

11.
0318. To act as pilot in command of an aircraft with a passenger aboard, the pilot must have made at least three takeoffs and three landings in an aircraft of the same category and class within the preceding

A— 90 days.
B— 120 days.
C— 12 months.
D— 24 months.

Answer (A) is correct (0318). *(FAR 61.57)*
To act as pilot in command of an airplane with passengers aboard, you must have made at least three takeoffs and landings (to a full stop if in a tailwheel airplane) in an airplane of the same category and class within the preceding 90 days. Category refers to airplane, helicopter, etc.; class refers to single or multiengine, land or sea.

12.
0320. The takeoffs and landings required to meet recency of experience requirements for carrying a passenger in a tailwheel airplane

A— may be touch-and-go in the same class airplane.
B— must be touch-and-go in any class airplane.
C— may be full stop in any class airplane.
D— must be full stop in the same class airplane.

Answer (D) is correct (0320). *(FAR 61.57)*
To comply with recency requirements for carrying a passenger in a tailwheel airplane, one must have made three takeoffs and landings to a full stop within the past 90 days. The requirement of three landings in 90 days also applies to tricycle gear airplanes, but the landings are not required to be full stop.
Answer (A) is incorrect because the landings may not be touch-and-go. Answer (B) is incorrect because they must be in the same class of airplane and they must be to a full stop. Answer (C) is incorrect because they must be in the same class of airplane. Class of airplane refers to single or multiengine, land or sea.

61.60 CHANGE OF ADDRESS

13.
0322. A certificated pilot changes permanent mailing address and fails to notify the FAA Airmen Certification Branch of the new address. For which period of time, after the move, may the pilot continue to exercise the privileges of his/her pilot certificate?

A— 30 days.
B— 60 days.
C— 90 days.
D— 120 days.

Answer (A) is correct (0322). *(FAR 61.60)*
If you have changed your permanent mailing address, you may not exercise the privileges of your pilot certificate after 30 days from the date of the address change unless you have notified the FAA of the change in writing. You are required to notify the

Airmen Certification Branch
Box 25082
Oklahoma City, OK 73125

61.101 RECREATIONAL PILOT PRIVILEGES AND LIMITATIONS

14.
0288. In regard to general privileges and limitations, a recreational pilot may

A— carry one passenger only.
B— carry more than one passenger.
C— carry up to four passengers.
D— not carry passengers.

Answer (A) is correct (0288). *(FAR 61.101)*
Recreational pilots may not carry more than one passenger.
Answer (B) is incorrect because <u>only</u> one passenger is permitted. Answer (C) is incorrect because although recreational pilots can fly an airplane certified for up to four occupants, there is a limit that only one passenger is permitted. Answer (D) is incorrect because one passenger is permitted.

Chapter 8: Federal Aviation Regulations

15.
0286. In regard to general privileges and limitations, a recreational pilot may

A— not be paid in any manner for the operating expenses of a flight.
B— fly for compensation or hire.
C— charge a reasonable fee for acting as pilot in command.
D— share the operating expenses of the flight with the passenger.

Answer (D) is correct (0286). *(FAR 61.101)*
Recreational pilots may share the operating expenses of the flight with the passenger.
Answer (A) is incorrect because sharing operating expense is being paid in some manner for the operating expenses. Answers (B) and (C) are incorrect because recreational pilots may not act as pilot-in-command of an aircraft for compensation or hire, nor may they act as pilot-in-command in furtherance of a business, nor if the aircraft is carrying a passenger or property for compensation or hire.

16.
0326. According to regulations pertaining to general privileges and limitations, a recreational pilot may

A— be paid for the operating expenses of a flight if at least three takeoffs and three landings were made by the pilot within the preceding 90 days.
B— not be paid in any manner for the operating expenses of a flight.
C— share the operating expenses of a flight with a passenger.
D— charge a reasonable fee for acting as pilot in command.

Answer (C) is correct (0326). *(FAR 61.101)*
Recreational pilots may share operating expenses of the flight with a passenger.
Answer (A) is incorrect because one cannot be paid for operating expenses; operating expenses can only be shared. Nonetheless, there is a requirement of three takeoffs and three landings within the preceding 90 days of when a passenger is to be carried. Answer (B) is incorrect because operating expenses can be shared, which in effect results in being compensated in some manner for part of the operating expenses. Answer (D) is incorrect because recreational pilots may not carry passengers or property for compensation or hire.

17.
0299. What is the maximum distance a recreational pilot may fly from the airport/heliport at which he/she received instruction?

A— 25 NM.
B— 50 NM.
C— 75 NM.
D— 100 NM.

Answer (B) is correct (0299). *(FAR 61.101)*
Recreational pilots may act as pilot-in-command of an aircraft when the flight is within 50 NM of an airport at which the pilot has received ground and flight instruction from an authorized instructor.
Answer (A) is incorrect because 25 NM is not a distance limitation for recreational pilots. Answers (C) and (D) are incorrect because recreational pilots cannot fly in excess of 50 NM away from the airport where they have received instruction.

18.
0289. A recreational pilot may act as pilot in command of an aircraft that is certificated for a maximum of

A— four occupants and has fixed landing gear.
B— four occupants and has retractable landing gear.
C— six occupants and has fixed landing gear.
D— six occupants and has retractable landing gear.

Answer (A) is correct (0289). *(FAR 61.101)*
Recreational pilots may not act as pilot-in-command in an aircraft certificated for more than four occupants nor in an aircraft with retractable landing gear. Thus recreational pilots can act as pilot-in-command of an airplane that is certificated for four occupants and fixed landing gear.
Answers (B) and (D) are incorrect because recreational pilots may not act as pilot-in-command of airplanes with retractable landing gear. Answers (C) and (D) are incorrect because recreational pilots cannot act as pilot-in-command of aircraft that are certificated for more than four occupants.

19.
0290. A recreational pilot may act as pilot in command of a properly certificated

A— 160 hp single-engine aircraft.
B— 180 hp multiengine airplane.
C— 200 hp single-engine airplane or helicopter.
D— 200 hp multiengine airplane.

Answer (A) is correct (0290). *(FAR 61.101)*
Recreational pilots may not act as pilot-in-command of an aircraft with a powerplant of more than 180 horsepower. Recreational pilots are also prohibited from operating aircraft with more than one engine.
Answers (B) and (D) are incorrect because recreational pilots can only operate single-engine aircraft. Answers (C) and (D) are incorrect because recreational pilots cannot operate aircraft with more than 180 horsepower.

20.
0287. What exception, if any, permits a recreational pilot to act as pilot in command of an aircraft carrying a passenger for hire?

A— If the passenger pays all the operating expenses only.
B— If a donation is made to a charitable organization for the flight.
C— If the pilot acts as second in command of an aircraft requiring more than one pilot.
D— There is no exception.

Answer (D) is correct (0287). *(FAR 61.101)*
A recreational pilot cannot act as pilot-in-command of an aircraft for compensation or hire. There are no exceptions. A recreational pilot may act as pilot-in-command of an airplane used in a passenger-carrying airlift sponsored by a charitable organization for which passengers make donations to the organization, provided the following requirements are met: the local GADO is notified, the flight is conducted from an adequate public airport, the pilot has logged at least 200 hrs, no acrobatic or formation flights are performed, the 100-hr inspection of the airplane requirement is complied with, and the flight is day-VFR.
Answer (A) is incorrect because recreational pilots may only share expenses with the passenger, i.e., the passenger cannot pay all the expenses. Answer (B) is incorrect because this is only permitted to private pilots. Answer (C) is incorrect because recreational pilots are precluded from acting as second-in-command when a second-in-command pilot is required.

21.
0291. May a recreational pilot act as pilot in command in furtherance of a business?

A— Yes, if the flight is only incidental to that business.
B— Yes, providing the aircraft does not carry a person or property for compensation or hire.
C— Yes, at any time.
D— No, it is not allowed.

Answer (D) is correct (0291). *(FAR 61.101)*
Recreational pilots may not act as pilot-in-command of aircraft in furtherance of a business. They also may not operate an aircraft for compensation or hire.
Answers (A), (B) and (C) are all incorrect because furtherance of <u>any</u> business is prohibited.

22.
0293. With respect to daylight hours, when is the earliest a recreational pilot may take off?

A— One hour before sunrise.
B— Thirty minutes before sunrise.
C— At sunrise.
D— At the beginning of morning civil twilight.

Answer (C) is correct (0293). *(FAR 61.101)*
Recreational pilots may not act as pilot-in-command of an aircraft between sunset and sunrise. Thus, recreational pilots may not take off before sunrise.
Answer (A) is incorrect because only private and higher-rated pilots may take off 1 hr before sunrise (i.e., night ends 1 hr before sunrise for them). Answer (B) is incorrect because there is no 30-minute leeway before sunrise or after sunset for recreational pilots. Answer (D) is incorrect because civil twilight is not mentioned in the recreational pilot FARs.

Chapter 8: Federal Aviation Regulations

23.
0292. If sunset is 2021 and the end of evening civil twilight is 2043, when must a recreational pilot terminate the flight?

A— 2021.
B— 2043.
C— 2051.
D— 2121.

24.
0295. When may a recreational pilot fly above 10,000 feet MSL?

A— When 2,000 feet AGL or below.
B— When 2,500 feet AGL or below.
C— When outside of controlled airspace.
D— Never.

25.
0296. During daytime, what is the minimum flight visibility required for recreational pilots flying in uncontrolled airspace and below 10,000 feet MSL?

A— 1 SM.
B— 3 SM.
C— 4 SM.
D— 5 SM.

26.
0298. During daytime, what is the minimum flight visibility required for recreational pilots in controlled airspace below 10,000 feet MSL?

A— 1 SM.
B— 3 SM.
C— 5 SM.
D— 7 SM.

27.
0377. What minimum visibility and clearance from clouds are required for a recreational pilot in uncontrolled airspace at 1,200 feet AGL or below during daylight hours?

A— 1 mile visibility and clear of clouds.
B— 1 mile visibility, 500 feet below the clouds.
C— 3 miles visibility and clear of clouds.
D— 3 miles visibility, 500 feet above the clouds.

Answer (A) is correct (0292). *(FAR 61.101)*
Recreational pilots may not act as pilot-in-command of an aircraft between sunset and sunrise; thus if sunset is 2021, they must terminate the flight by 2021.

Answer (B) is incorrect because civil twilight has nothing to do with the sunset rules in the FARs. Answer (C) is incorrect because there is no 30-minute leeway after sunset for recreational pilots. Answer (D) is incorrect because 2121 is when night flight begins for private and higher-rated pilots per FAR 61.57 (but not for recreational pilots).

Answer (A) is correct (0295). *(FAR 61.101)*
Recreational pilots may not act as pilot-in-command of an aircraft above 10,000 ft MSL or 2,000 ft AGL, whichever is higher.

Answer (B) is incorrect because one is not permitted to be above 2,000 ft AGL. Answer (C) is incorrect because the FAR specifies 10,000 ft MSL with no distinction as to controlled or uncontrolled. Answer (D) is incorrect because recreational pilots can fly above 10,000 ft MSL as long as they are not above 2,000 ft AGL.

Answer (B) is correct (0296). *(FAR 61.101)*
Recreational pilots are not permitted to act as pilot-in-command when the flight or surface visibility is less than 3 statute miles, whether in controlled or uncontrolled airspace.

Answer (A) is incorrect because recreational pilots can never fly when the visibility is less than 3 miles. Answer (C) is incorrect because 4 miles of visibility is not a requirement for any pilot. Answer (D) is incorrect because 5 statute miles minimum visibility is the <u>night</u> requirement for recreational pilots.

Answer (B) is correct (0298). *(FAR 61.101)*
Recreational pilots are not permitted to act as pilot-in-command when the flight or surface visibility is less than 3 statute miles, whether in controlled or uncontrolled airspace.

Answer (A) is incorrect because recreational pilots can never fly when the visibility is less than 3 miles. Answer (C) is incorrect because 5 statute miles minimum visibility is the <u>night</u> requirement for recreational pilots. Answer (D) is incorrect because 7 miles of visibility is not a requirement for any pilot.

Answer (C) is correct (0377). *(FARs 61.101, 91.105)*
FAR 61.101 requires that the minimum visibility during the day for recreational pilots is 3 miles. The recreational pilot FARs do not have any regulations different from other VFR pilots regarding distance from clouds. FAR 91.105 requires pilots to remain clear of clouds when below 1,200 ft AGL.

Answers (A) and (B) are incorrect because recreational pilots may not fly in less than 3 miles visibility during the day. Answers (B) and (D) are incorrect because recreational and other pilots need only remain clear of clouds when below 1,200 ft AGL.

28.

0297. When may a recreational pilot fly above a ceiling?

A— Anytime, except over an obscuration.
B— When there are breaks in an overcast layer.
C— When maintaining visual reference with the surface.
D— Never.

29.

0300. When, if ever, may a recreational pilot act as pilot in command in an aircraft towing a banner?

A— If the pilot has logged 100 hours of flight time in a powered aircraft.
B— If the pilot has an endorsement in his/her pilot logbook from an authorized flight instructor.
C— Anytime, unless it is for compensation or hire.
D— It is not allowed.

30.

0294. When may a recreational pilot fly as sole occupant of an aircraft at night?

A— While under the supervision of a flight instructor.
B— While under the supervision of a flight instructor provided the flight or surface visibility is at least 3 SM.
C— While under the supervision of a flight instructor provided the flight or surface visibility is at least 5 SM.
D— Never.

Answer (C) is correct (0297). *(FAR 61.101)*
Recreational pilots are not permitted to act as pilot-in-command of an aircraft without visual reference to the surface. A ceiling includes a broken layer of clouds covering 50% or more of the sky which, when near 50%, may permit visual reference to the surface.
Answer (A) is incorrect because recreational pilots may only fly above a ceiling when there is visual reference to the surface. Answer (B) is incorrect because the issue is not whether or not there are breaks in the overcast layers, but rather whether the pilot can maintain visual reference with the surface. Answer (D) is incorrect because recreational pilots are permitted to fly above the ceiling, e.g., broken clouds, if they can maintain visual reference to the ground.

Answer (D) is correct (0300). *(FAR 61.101)*
Recreational pilots may not act as pilot-in-command of an aircraft that is towing any object.
Answer (A) is incorrect because 100 hrs of flight time relates to glider-towing experience requirements if a private or higher-rated pilot towed 10 gliders prior to May 17, 1967. Answer (B) is incorrect because a flight instructor endorsement is one of the requirements for glider-towing by private or higher-rated pilot. Answer (C) is incorrect because towing <u>any</u> objects by recreational pilots is in violation of FAR 61.101.

Answer (C) is correct (0294). *(FAR 61.101)*
For the purpose of obtaining a private pilot certificate, a recreational pilot may fly as sole occupant of an aircraft between sunset and sunrise provided that the flight or surface visibility is at least 5 statute miles and under the supervision of a CFI.
Answers (A) and (B) are incorrect because FAR 61.101 specifically requires 5 statute miles visibility at night. Answer (D) is incorrect because the FARs provide a mechanism for pilots to upgrade from the recreational pilot certificate to the private pilot certificate, which includes gaining night experience.

Chapter 8: Federal Aviation Regulations

31.
0301. When must a recreational pilot have a pilot in command flight check?

A— Every 400 hours.
B— Every 180 days.
C— If the pilot has less than 400 total flight hours and has not flown as pilot in command in an aircraft within the preceding 180 days.
D— If the pilot has less than 500 total flight hours and has not flown as pilot in command in an aircraft within the preceding 90 days.

Answer (C) is correct (0301). *(FAR 61.101)*
In addition to an annual flight review, a pilot-in-command flight check is required of recreational pilots with less than 400 hrs of flight time if they have not flown within the preceding 180 days. The pilot-in-command check requires flight instruction from a CFI who must endorse the recreational pilot's logbook that the pilot is competent to act as pilot-in-command of the aircraft. This check can also constitute the annual flight review required by FAR 61.56, if so endorsed.
Answer (A) is incorrect because recreational pilots with 400 or more flight hrs are not subject to the pilot-in-command flight check. Answer (B) is incorrect because if recreational pilots fly on a regular basis and do not skip 180 days without flying, no pilot-in-command flight check is required. Answer (D) is incorrect because recreational pilots with 400 or more flight hrs are not subject to the pilot-in-command flight check. The 90 days refers to the three-takeoffs-and-landings requirement before carrying a passenger.

71.5 EXTENT OF FEDERAL AIRWAYS

32.
0323. Unless otherwise specified, Federal airways extend from

A— 700 feet above the surface upward to the Continental Control Area and are 10 NM wide.
B— 1,200 feet above the surface upward to 14,500 feet MSL and are 16 NM wide.
C— 1,200 feet above the surface upward to 18,000 feet MSL and are 8 NM wide.
D— the surface upward to 18,000 feet MSL and are 4 NM wide.

Answer (C) is correct (0323). *(FAR 71.5)*
Federal airways include airspace extending upward from 1,200 ft AGL to, but not including, 18,000 ft MSL. They also include airspace 4 miles each side of the center line of the airway; hence, they are 8 miles wide.
Answers (A) and (B) are incorrect because airways extend up to 18,000 ft, not 14,500 ft (the base of the Continental Control Area). Answers (A), (B), and (D) are incorrect because the airways are 8 NM wide. Answers (A) and (D) are incorrect because airways begin at 1,200 ft AGL.

71.11 CONTROL ZONES

33.
0324. Within the contiguous United States, a control zone extends from the surface upward to the base of

A— an Airport Traffic Area.
B— the Continental Control Area.
C— a Transition Area.
D— a Terminal Control Area.

Answer (B) is correct (0324). *(FAR 71.11)*
Control zones extend upward from the surface to the base of the Continental Control Area, or 14,500 ft MSL. The Continental Control Area exists over the 48 contiguous United States, the District of Columbia, and most of Alaska.
Answer (A) is incorrect because it is a circular area within a 5-mile radius of airports with operating control towers. Answer (C) is incorrect because a transition area concerns IFR approach routes near airports for the purpose of making them controlled airspace. Answer (D) is incorrect because a terminal control area is controlled airspace in the general vicinity of high-density air traffic.

91.3 RESPONSIBILITY AND AUTHORITY OF PILOT-IN-COMMAND

34.
0325. If an in-flight emergency requires immediate action, a pilot in command may

A— deviate from FAR's to the extent required to meet that emergency.
B— deviate from FAR's to the extent required to meet the emergency, but must submit a written report to the Administrator within 24 hours.
C— not deviate from FAR's unless permission is obtained from Air Traffic Control.
D— not deviate from FAR's unless prior to the deviation approval is granted by the Administrator.

Answer (A) is correct (0325). *(FAR 91.3)*
In an emergency, the pilot-in-command may deviate from any regulation as necessary to meet that emergency. Any pilot who makes such a deviation must submit a written report to the FAA if requested by ATC.
Answer (B) is incorrect because the written report should only be submitted if requested by ATC, and if so, there is no 24-hr deadline. Answer (C) is incorrect because the deviation can be immediate if the emergency is immediate, i.e., without ATC permission. Answer (D) is incorrect because it describes a waiver from the FARs which does not imply an emergency.

35.
0327. The final authority as to the operation of an aircraft is the

A— FAA.
B— pilot in command.
C— owner or operator of the aircraft.
D— inspector who performs the periodic check.

Answer (B) is correct (0327). *(FAR 91.3)*
FAR 91.3 states, "The pilot-in-command of an aircraft is directly responsible for, and is the final authority as to, the operation of that aircraft."

91.5 PREFLIGHT ACTION

36.
0331. Prior to each flight, the pilot in command must

A— check the personal logbook for appropriate recent experience.
B— become familiar with all available information concerning that flight.
C— calculate the weight and balance of the aircraft to determine if the CG is within limits.
D— check with ATC for the latest traffic advisories and any possible delays.

Answer (B) is correct (0331). *(FAR 91.5)*
The FARs require each pilot-in-command to be familiar with all available information concerning the flight before beginning any flight.
Answers (A), (C), and (D) are incorrect because these actions may be advisable on all flights, but are not required in every circumstance.

37.
0328. Which preflight action is required for every flight?

A— Check weather reports and forecasts.
B— Determine runway length at airports of intended use.
C— Determine alternatives if the flight cannot be completed.
D— Check for any known traffic delays.

Answer (B) is correct (0328). *(FAR 91.5)*
The FARs require each pilot-in-command to be familiar with all available information concerning the flight before beginning any flight. A specific requirement is that runway lengths at airports of intended use be determined.
Answers (A), (C), and (D) are incorrect because each is required only for flights not in the vicinity of an airport or under IFR. They are pertinent for other flights but not required.

Chapter 8: Federal Aviation Regulations

38.
0330. In addition to other preflight actions for a VFR flight away from the vicinity of the departure airport, regulations require the pilot in command to

A— file a flight plan.
B— check each fuel tank visually to ensure that it is full.
C— check the accuracy of the omninavigation equipment and the emergency locator transmitter.
D— determine runway lengths of airports of intended use and the airplane's takeoff and landing distance data.

Answer (D) is correct (0330). *(FAR 91.5)*
For all flights, runway lengths at airports of intended use must be determined before the flight is begun.
Answers (A), (B), and (C) are incorrect because, while each is advisable, they are not required by the FARs.

39.
0329. Preflight action, as required by regulations for all flights away from the vicinity of an airport, shall include a study of the weather, taking into consideration fuel requirements and

A— an operational check of the navigation radios.
B— the designation of an alternate airport.
C— the filing of a flight plan.
D— an alternate course of action if the flight cannot be completed as planned.

Answer (D) is correct (0329). *(FAR 91.5)*
Preflight actions for flights not in the vicinity of an airport include checking weather reports and forecasts, fuel requirements, alternatives available if the planned flight cannot be completed, and any known traffic delays.
Answers (A), (B), and (C) are incorrect because each is advisable and is generally required for IFR, but they are not required for VFR.

91.11 ALCOHOL OR DRUGS

40.
0332. A person may not act as a crewmember of a civil aircraft if alcoholic beverages have been consumed by that person within the preceding

A— 8 hours.
B— 12 hours.
C— 24 hours.
D— 48 hours.

Answer (A) is correct (0332). *(FAR 91.11)*
Regulations require that no person may act as a crewmember (including pilot-in-command) of a civil aircraft while

(1) using any drug that affects his/her faculties in any way contrary to safety, or
(2) within 8 hrs after the consumption of any alcoholic beverages.

91.14 USE OF SAFETY BELTS

41.
0333. What obligation does a pilot in command have concerning the use of seatbelts?

A— The pilot in command's seatbelt must be fastened during the entire flight.
B— The pilot in command must brief the passenger on the use of the seatbelt and notify him/her to fasten the seatbelt during takeoff and landing.
C— The pilot in command must instruct the passenger to keep the seatbelt fastened for the entire flight except for brief rest periods.
D— The pilot in command must instruct the passenger on the use of the seatbelt, but does not have the authority to require their use, except during an emergency.

Answer (B) is correct (0333). *(FAR 91.14)*
Before taking off or landing, pilots must advise the passengers on how to use the seatbelts. Pilots also must notify them to fasten their seatbelts for takeoff and landing.
Answers (A) and (C) are incorrect because the seatbelt requirement applies only to landings and takeoffs for both crewmembers and passengers. Answer (D) is incorrect because the pilot-in-command has the responsibility both to instruct passengers on the use of seatbelts and to require their use on takeoffs and landings.

91.15 PARACHUTES AND PARACHUTING

42.
0334. When must each occupant of an aircraft wear an approved parachute?

A— When an aircraft is being tested after major repair.
B— When flying over water beyond gliding distance to the shore.
C— When a door is removed from the aircraft to facilitate parachute jumpers.
D— When an intentional bank that exceeds 60° is to be made.

Answer (D) is correct (0334). *(FAR 91.15)*
Unless each occupant of an airplane is wearing an approved parachute, no pilot carrying any other person or crewmember may execute any intentional maneuver that exceeds a bank of 60° or a noseup or nosedown attitude of 30° relative to the horizon.
Answer (A) is incorrect because the parachute restriction does not apply to flight tests. Answer (B) is incorrect because approved flotation devices, not parachutes, must be carried in the aircraft over water. Answer (C) is incorrect because pilots of airplanes that are carrying parachute jumpers are not required to use a parachute.

43.
0336. An approved chair-type parachute may be carried for emergency use if it has been packed by an appropriately rated parachute rigger within the preceding

A— 60 days.
B— 90 days.
C— 120 days.
D— 6 months.

Answer (C) is correct (0336). *(FAR 91.15)*
"No pilot of a civil aircraft may allow a parachute that is available for emergency use to be carried unless it is an approved type and, if a chair type (canopy in back), it has been packed by a certificated and appropriately rated parachute rigger within the preceding 120 days."

44.
0335. A chair-type parachute must have been packed by a certificated and appropriately rated parachute rigger within the preceding

A— 30 days.
B— 60 days.
C— 90 days.
D— 120 days.

Answer (D) is correct (0335). *(FAR 91.15)*
"No pilot of a civil aircraft may allow a parachute that is available for emergency use to be carried unless it is an approved type and, if a chair type (canopy in back), it has been packed by a certificated and appropriately rated parachute rigger within the preceding 120 days."

91.22 FUEL REQUIREMENTS FOR FLIGHT UNDER VFR

45.
0339. What is the fuel requirement for flight under VFR during daylight hours?

A— Full fuel tanks.
B— Enough to complete the flight at normal cruising flight with adverse wind conditions.
C— Enough to fly to the first point of intended landing, and assuming normal cruising speed, fly thereafter for at least 30 minutes.
D— Enough to fly to the first point of intended landing, and assuming normal cruising speed, fly thereafter for at least 45 minutes.

Answer (C) is correct (0339). *(FAR 91.22)*
The day-VFR requirement is enough fuel to fly to the first point of intended landing and thereafter for 30 minutes at normal cruising speed.
Answer (A) is incorrect because full fuel tanks may put the plane over gross weight. Answer (B) is incorrect because the fuel requirements are based upon the wind conditions existing that day plus the 30-minute reserve. Answer (D) is incorrect because a 45-minute reserve is the requirement for night flight.

Chapter 8: Federal Aviation Regulations

91.27 CIVIL AIRCRAFT CERTIFICATIONS REQUIRED

46.
0340. What documents or records must be aboard an aircraft during flight?

A— Operating limitations and an aircraft use and inspection report.
B— Operating limitations; a Registration Certificate; and an appropriate, current, and properly displayed Airworthiness Certificate.
C— Repair and alteration forms and a Registration Certificate.
D— Aircraft and engine logbooks and a Registration Certificate.

Answer (B) is incorrect (0340). *(FARs 91.27, 91.31)*
All civil aircraft must have an appropriate and current airworthiness certificate, as well as a registration certificate, on board. The airworthiness certificate should be displayed at the cabin or cockpit entrance so that it is visible to passengers or crew. FAR 91.31 requires that operating limitations in the form of flight manuals, markings, and/or placards be available in the aircraft.
Answers (A), (C), and (D) are incorrect because they omit the airworthiness certificate. Answer (A) also omits the registration certificate. Answers (C) and (D) also omit the operating limitations.

47.
0341. No person may operate a civil aircraft unless the Airworthiness Certificate or special flight permit or authorization required by regulations is

A— on file in the owner's operation office where the aircraft is based.
B— filed with the other required certificates or documents within the aircraft to be flown.
C— displayed at the cabin or cockpit entrance so that it is legible to the passenger or crewmember.
D— included in the approved logbooks for the aircraft to be flown.

Answer (C) is correct (0341). *(FAR 91.27)*
All civil aircraft must have an appropriate and current airworthiness certificate, as well as a registration certificate, on board. The airworthiness certificate should be displayed at the cabin or cockpit entrance so that it is visible to passengers or crew.
Answers (A), (B), and (D) are incorrect because the airworthiness certificate must be displayed in the cabin or cockpit, not filed with records or logbooks.

91.29 CIVIL AIRCRAFT AIRWORTHINESS

48.
0342. Who is responsible for determining if an aircraft is in condition for safe flight?

A— A certificated aircraft mechanic.
B— A certificated aircraft maintenance inspector.
C— The pilot in command.
D— The owner or operator.

Answer (C) is correct (0342). *(FAR 91.29)*
The pilot-in-command of an aircraft is directly responsible for, and is the final authority for, determining whether the airplane is in safe condition.
Answers (A), (B), and (D) are incorrect because the responsibilities of mechanics, inspectors, and owners do not relieve each pilot of the responsibility to determine that the airplane is airworthy prior to every flight.

91.31 CIVIL AIRCRAFT FLIGHT MANUAL, MARKING, AND PLACARD REQUIREMENTS

49.
0343. An aircraft's operating limitations may be found

A— on the Airworthiness Certificate.
B— in the airplane flight manual, approved manual material, markings, and placards.
C— only in the FAA-approved airplane flight manual.
D— in the manufacturer's Airworthiness Directives.

Answer (B) is correct (0343). *(FAR 91.31)*
An airplane's operating limitations may be found in the airplane flight manual, in approved flight manual material, markings, and placards, or any combination thereof.
Answer (A) is incorrect because the airworthiness certificate only indicates the airplane was in an airworthy condition when delivered from the factory. Answer (C) is incorrect because operating limitations may also be present in markings and placards within the airplane. Answer (D) is incorrect because airworthiness directives are issued by the FAA, not the manufacturer, and involve mandatory repairs, alterations, etc.

91.39 RESTRICTED CATEGORY CIVIL AIRCRAFT; OPERATING LIMITATIONS

50.
0344. Which is normally prohibited when operating a restricted category civil aircraft?

A— Flight under instrument flight rules.
B— Flight over a densely populated area.
C— Flight within a control zone.
D— Flight within the Continental Control Area.

Answer (B) is correct (0344). *(FAR 91.42)*
No person may operate an aircraft that has an experimental certificate over a densely populated area or in a congested airway unless authorized by the administrator.

91.52 EMERGENCY LOCATOR TRANSMITTERS

51.
0345. When are non-rechargeable batteries of an emergency locator transmitter (ELT) required to be replaced?

A— Every 24 months.
B— When 50 percent of their useful life expires or they were in use for a cumulative period of 1 hour.
C— At the time of each 100-hour or annual inspection.
D— Annually.

Answer (B) is correct (0345). *(FAR 91.52)*
Emergency locator transmitter batteries must be replaced or recharged after 50% of their useful life, as established by the transmitter manufacturer or one cumulative hour of use.

91.67 RIGHT-OF-WAY RULES

52.
0350. The aircraft which has the right-of-way over all other air traffic is an aircraft

A— towing a banner.
B— in distress.
C— on final approach to land.
D— towing or refueling another aircraft.

Answer (B) is correct (0350). *(FAR 91.67)*
An aircraft in distress has the right-of-way over all other aircraft.
Answer (A) is incorrect because no special right-of-way provisions are granted to aircraft towing banners. Answer (C) is incorrect because aircraft on the final approach of a landing have the right-of-way over other aircraft in flight or operating on the surface. Answer (D) is incorrect because aircraft towing or refueling another aircraft only have the right-of-way over other engine-driven aircraft.

53.
0349. What action should be taken if a glider and an airplane approach each other at the same altitude and on a head-on collision course?

A— The airplane should give way because the glider has the right-of-way.
B— The airplane should give way because it is more maneuverable.
C— Both should give way to the right.
D— The glider should descend and the airplane should climb so as to pass each other by at least 500 feet.

Answer (C) is correct (0349). *(FAR 91.67)*
When aircraft are approaching head-on, each aircraft shall alter course to the right.
Answers (A) and (B) are incorrect because the glider has the right-of-way only if the aircraft are converging, but not head-on. Answer (D) is incorrect because the requirement is for both planes to alter course to the right regardless of category.

Chapter 8: Federal Aviation Regulations

54.
0351. Which aircraft has the right-of-way over all other aircraft?

A— Balloon.
B— Airship.
C— Rotorcraft.
D— Airplane.

Answer (A) is correct (0351). *(FAR 91.67)*
If aircraft of different categories are converging, the right-of-way depends upon who has the least maneuverability. A balloon has the right-of-way over all other aircraft. A glider has right-of-way over an airship, airplane or rotorcraft. An airship has right-of-way over an airplane or rotorcraft.

Answer (B) is incorrect because an airship has right-of-way over an airplane or rotorcraft. Answers (C) and (D) are incorrect because airplanes and rotorcrafts do not have right-of-way over other aircraft, per se.

55.
0346. An airplane and an airship are converging. If the airship is left of the airplane's position, which aircraft has the right-of-way?

A— The pilot of the airplane should give way; the airship is to the left.
B— The airship has the right-of-way.
C— Each pilot should alter course to the right.
D— The airplane has the right-of-way; it is more maneuverable.

Answer (B) is correct (0346). *(FAR 91.67)*
When aircraft of different categories are converging, airships have the right-of-way over airplanes or rotorcraft. Thus, the airship has the right-of-way in this question.

Answer (A) is incorrect because the airship is less maneuverable and has the right-of-way when converging with an airplane. Answer (C) is incorrect because it is the avoidance technique if approaching head on. Answer (D) is incorrect because the less maneuverable aircraft (the airship in this instance) usually has the right-of-way.

56.
0348. What action is required when two aircraft of the same category converge at the same altitude, but not head-on?

A— The more maneuverable aircraft shall give way.
B— The faster aircraft shall give way.
C— The aircraft on the left shall give way.
D— Each aircraft shall give way to the right.

Answer (C) is correct (0348). *(FAR 91.67)*
When two aircraft of the same category converge (but not head on), the aircraft to the other's right has the right-of-way. Therefore, an airplane on the left gives way to the airplane on the right.

57.
0347. When two or more airplanes are approaching an airport for the purpose of landing, the right-of-way belongs to the airplane

A— that has the other to its right.
B— that is the least maneuverable.
C— that is either ahead of or to the other's right regardless of altitude.
D— at the lower altitude, but it shall not take advantage of this rule to cut in front of or to overtake another.

Answer (D) is correct (0347). *(FAR 91.67)*
When two or more airplanes are approaching an airport for the purpose of landing, the airplane at the lower altitude has the right-of-way, but an airplane flying at a higher altitude shall not take advantage of this rule to cut in front of or to overtake another airplane.

91.70 AIRCRAFT SPEED

58.
0353. Unless otherwise authorized or required by Air Traffic Control, what is the maximum indicated airspeed at which a person may operate an aircraft below 10,000 feet MSL?

A— 156 knots.
B— 180 knots.
C— 200 knots.
D— 250 knots.

Answer (D) is correct (0353). *(FAR 91.70)*
Unless authorized, no person may operate an aircraft below 10,000 ft MSL at an indicated airspeed of more than 250 kts (288 mph). If the aircraft cannot fly at that speed safely, the pilot may fly at whatever is the minimum safe speed for the aircraft.

59.
0352. When flying beneath the lateral limits of a Terminal Control Area, the maximum speed authorized is

A— 156 knots.
B— 180 knots.
C— 200 knots.
D— 250 knots.

Answer (C) is correct (0352). *(FAR 91.70)*
No person may operate an airplane in the airspace beneath a terminal control area or in a VFR corridor designated through a TCA at an indicated airspeed of more than 200 kts (230 mph).

91.71 ACROBATIC FLIGHT

60.
0356. No person may operate an aircraft in acrobatic flight when the flight visibility is less than

A— 3 miles.
B— 5 miles.
C— 7 miles.
D— 10 miles.

Answer (A) is correct (0356). *(FAR 91.71)*
Acrobatic flight is not permitted when visibility is less than 3 miles.

61.
0357. What is the lowest altitude permitted for acrobatic flight?

A— 1,000 feet AGL.
B— 1,000 feet above the highest obstacle within 5 miles.
C— 1,500 feet AGL.
D— 1,500 feet above the highest obstacle within 5 miles.

Answer (C) is correct (0357). *(FAR 91.71)*
Acrobatic flight is not permitted below 1,500 ft AGL.

62.
0354. In which controlled airspace is acrobatic flight prohibited?

A— All controlled airspace.
B— Control zones and Federal airways.
C— Control zones, Federal airways, and control areas.
D— Control zones, Federal airways, control areas, and Traffic Control Areas.

Answer (B) is correct (0354). *(FAR 91.71)*
Acrobatic flight may not be undertaken over any congested area of a city, town, or settlement; over an open-air assembly of persons; within a control zone or federal airway; below an altitude of 1,500 ft AGL; or when flight visibility is less than 3 miles.
Answer (A) is incorrect because acrobatic flight is only prohibited from control zones and federal airways. Answers (C) and (D) are incorrect because acrobatic flight is not prohibited in control areas.

Chapter 8: Federal Aviation Regulations

63.
0355. According to regulations, no person may operate an aircraft in acrobatic flight

A— over any congested area of a city, town, or settlement.
B— within 5 miles of a Federal airway.
C— below an altitude of 2,000 feet above the surface.
D— when flight visibility is less than 5 miles.

Answer (A) is correct (0355). *(FAR 91.71)*
No person may operate an aircraft in acrobatic flight over an open air assembly of people; over a congested area of a city, town, or settlement; within a control zone or federal airway; below 1,500 ft AGL in less than 3 miles visibility.
Answer (B) is incorrect because the regulation does not allow acrobatics in a federal airway. Answer (C) is incorrect because the minimum altitude is 1,500 ft AGL, not 2,000 ft AGL. Answer (D) is incorrect because the minimum visibility is 3 miles, not 5 miles.

91.79 MINIMUM SAFE ALTITUDES

64.
0360. Except when necessary for takeoff or landing, what is the minimum safe altitude required for a pilot to operate an aircraft over congested areas?

A— An altitude allowing, if a power unit fails, an emergency landing without undue hazard to persons or property on the surface.
B— An altitude of 500 feet above the surface and no closer than 500 feet to any person, vessel, vehicle, or structure.
C— An altitude of 500 feet above the highest obstacle within a horizontal radius of 1,000 feet.
D— An altitude of 1,000 feet above the highest obstacle within a horizontal radius of 2,000 feet.

Answer (D) is correct (0360). *(FAR 91.79)*
When operating an aircraft over a congested area, the pilot must remain at least 1,000 ft above the highest obstacle within a horizontal radius of 2,000 ft of the aircraft. In other words, the airplane must be 1,000 ft above a 2,000-ft radius about the top of the highest obstacle.
Answer (A) is incorrect because this is a general requirement (not specific to congested areas). Answer (B) is incorrect because, in sparsely populated areas, one must keep 500 ft from any person, vessel, vehicle, or structure. Answer (C) is incorrect because the 500 ft AGL rule has no horizontal radius requirements.

65.
0361. Except when necessary for takeoff or landing, what is the minimum safe altitude for a pilot to operate an airplane over other than a congested area?

A— An altitude allowing, if a power unit fails, an emergency landing without undue hazard to persons or property on the surface.
B— An altitude of 500 feet AGL except over open water or a sparsely populated area which requires 500 feet from any person, vessel, vehicle, or structure.
C— An altitude of 500 feet above the highest obstacle within a horizontal radius of 1,000 feet.
D— An altitude of 1,000 feet above the highest obstacle within a horizontal radius of 2,000 feet.

Answer (B) is correct (0361). *(FAR 91.79)*
Over other than congested areas, an altitude of 500 ft above the surface is required. Over open water and sparsely populated areas, a distance of 500 ft from any person, vessel, vehicle, or structure must be maintained.

Chapter 8: Federal Aviation Regulations

66.
0358. Except when necessary for takeoff or landing, an aircraft may not be operated closer than what distance from any person, vehicle, or structure?

A— 500 feet.
B— 1,000 feet.
C— 1,500 feet.
D— 2,000 feet.

Answer (A) is correct (0358). *(FAR 91.79)*
Over other than congested areas, an altitude of 500 ft above the surface is required. Over open water and sparsely populated areas, a distance of 500 ft from any person, vessel, vehicle, or structure must be maintained.

67.
0359. Except when necessary for takeoff or landing, what is the minimum safe altitude for a pilot to operate an aircraft anywhere?

A— An altitude allowing, if a power unit fails, an emergency landing without undue hazard to persons or property on the surface.
B— An altitude of 500 feet above the surface and no closer than 500 feet to any person, vessel, vehicle, or structure.
C— An altitude of 500 feet above the highest obstacle within a horizontal radius of 1,000 feet.
D— An altitude of 1,000 feet above the highest obstacle within a horizontal radius of 2,000 feet.

Answer (A) is correct (0359). *(FAR 91.79)*
FAR 91.79 provides for minimum safe altitudes anywhere, over congested areas, and over other than congested areas. In the anywhere category, the altitude must be sufficient to allow for an emergency landing without undue hazard to persons or property on the surface if a power unit fails. There is an exemption from all altitude requirements for takeoff and landing.

91.90 TERMINAL CONTROL AREAS

68.
0369. What minimum pilot certification is required for operations in a Terminal Control Area?

A— Recreational Pilot Certificate.
B— Private Pilot Certificate.
C— Private Pilot Certificate with an instrument rating.
D— Commercial Pilot Certificate.

Answer (B) is correct (0369). *(FAR 91.90)*
A pilot must have at least a private pilot certificate for operating in a TCA. Student pilots and recreational pilots after special instruction and logbook endorsements (FAR 61.95) can operate in a TCA.
Answer (A) is incorrect because recreational pilots are not permitted to act as pilot-in-command in airspace requiring ATC authorization. Answers (C) and (D) are incorrect because only a private pilot certificate is required.

91.95 RESTRICTED AND PROHIBITED AREAS

69.
0272. Under what condition, if any, may civil pilots enter a restricted area?

A— For takeoff and landing to take care of official business.
B— With the controlling agency's authorization.
C— On airways with ATC clearance.
D— Under no condition.

Answer (B) is correct (0272). *(FAR 91.95)*
An aircraft may not be operated within a restricted area unless permission has been obtained from the controlling agency. Frequently, the ATC within the area acts as the controlling agent's authorization; e.g., ATC in a restricted military area can permit aircraft to enter when the restricted area is not active.
Answer (A) is incorrect because authorization must be obtained. Answer (C) is incorrect because airways do not penetrate restricted areas. Answer (D) is incorrect because restricted areas may be entered with proper authorization.

Chapter 8: Federal Aviation Regulations 165

91.105 BASIC VFR WEATHER MINIMUMS

70.
0378. During operations within controlled airspace at altitudes of more than 1,200 feet AGL, but less than 10,000 feet MSL, the minimum distance below clouds requirement for VFR flight is

A— 500 feet.
B— 1,000 feet.
C— 1,500 feet.
D— 2,000 feet.

Answer (A) is correct (0378). *(FAR 91.105)*
When operating within controlled airspace at altitudes of more than 1,200 ft AGL but less than 10,000 ft MSL, the minimum distance below clouds is 500 ft.
Answer (B) is incorrect because 1,000 ft is the above-cloud requirement. Answer (C) is incorrect because 1,500 ft is not a cloud separation. Answer (D) is incorrect because 2,000 ft is the required horizontal separation.

71.
0382. During operations within controlled airspace at altitudes of less than 1,200 feet AGL, the minimum horizontal distance from clouds requirement for VFR flight is

A— 500 feet.
B— 1,000 feet.
C— 1,500 feet.
D— 2,000 feet.

Answer (D) is correct (0382). *(FAR 91.105)*
When in controlled airspace at less than 1,200 ft AGL, the minimum horizontal distance from clouds is 2,000 ft. In uncontrolled airspace, the requirement is to remain clear of clouds.

72.
0374. The basic VFR weather minimums for operating an aircraft beneath the ceiling within a control zone are

A— 500-foot ceiling and 1 mile visibility.
B— 1,000-foot ceiling and 3 miles visibility.
C— clear of clouds and 2 miles visibility.
D— 2,000-foot ceiling and 1 mile visibility.

Answer (B) is correct (0374). *(FAR 91.105)*
Since a control zone is controlled airspace, 3 statute miles visibility is required for VFR flight per 91.105. FAR 91.101 requires a minimum of 3 miles visibility for recreational pilots in both controlled and uncontrolled airspace. Except with a special clearance (not available to recreational pilots), no one may operate under VFR in a control zone, beneath the cloud ceiling unless the ceiling is 1,000 ft or more.

73.
0380. Outside controlled airspace, the minimum flight visibility requirement for a recreational pilot flying VFR above 1,200 feet AGL and below 10,000 feet MSL during daylight hours is

A— 1 mile.
B— 3 miles.
C— 5 miles.
D— 7 miles.

Answer (B) is correct (0380). *(FAR 91.101)*
Recreational pilots are required to have 3 miles visibility whether in controlled or uncontrolled airspace during the day and 5 miles visibility at night (the latter is only during solo under the supervision of a CFI).

74.
0381. During operations outside controlled airspace at altitudes of more than 1,200 feet AGL, but less than 10,000 feet MSL, the minimum distance below clouds requirement for VFR flight is

A— 500 feet.
B— 1,000 feet.
C— 1,500 feet.
D— 2,000 feet.

Answer (A) is correct (0381). *(FAR 91.105)*
The distance from clouds requirement for operations above 1,200 ft AGL in uncontrolled airspace is the same as for controlled airspace: 500 ft below, 1,000 ft above, and 2,000 ft horizontally.

75.
0371. During operations outside controlled airspace at altitudes of more than 1,200 feet AGL, but less than 10,000 feet MSL, the minimum horizontal distance from clouds requirement for VFR flight is

A— 500 feet.
B— 1,000 feet.
C— 1,500 feet.
D— 2,000 feet.

Answer (D) is correct (0371). *(FAR 91.105)*
Above 1,200 ft AGL but less than 10,000 ft MSL, the minimum distance over clouds is 1,000 ft. The requirement under clouds is 500 ft and the horizontal separation is 2,000 ft. This is true whether in or out of controlled airspace.

76.
0379. During operations at altitudes of more than 1,200 feet AGL and at or above 10,000 feet MSL, the minimum distance below clouds requirement for VFR flight is

A— 500 feet.
B— 1,000 feet.
C— 1,500 feet.
D— 2,000 feet.

Answer (B) is correct (0379). *(FAR 91.105)*
Above 1,200 ft AGL and above 10,000 ft MSL, the minimum distance below clouds is 1,000 ft. This requirement is the same for controlled and uncontrolled airspace. The required clearance is 1,000 feet above and below clouds. Under 10,000 ft MSL, VFR flight may be only 500 ft below clouds.

77.
0376. What minimum flight visibility is required for VFR flight operations on an airway below 10,000 feet?

A— 1 mile.
B— 3 miles.
C— 4 miles.
D— 5 miles.

Answer (B) is correct (0376). *(FAR 91.105)*
Since airways are controlled airspace, the VFR minimum visibility below 10,000 ft MSL is 3 statute miles.

78.
0372. VFR flight above 1,200 feet AGL and below 10,000 feet MSL requires a minimum visibility and vertical cloud clearance of

A— 3 miles, and 500 feet below the clouds.
B— 3 miles, and 1,000 feet below the clouds.
C— 5 miles, and 500 feet below the clouds.
D— 5 miles, and 1,000 feet below the clouds.

Answer (A) is correct (0372). *(FAR 91.105)*
When above 1,200 ft AGL but below 10,000 ft MSL, the minimum visibility is 3 statute miles. Clearance must be 500 ft below and 1,000 ft above clouds. The latter is true whether in or out of controlled airspace.

A memory aid is to be higher (1,000 ft) above and lower (500 ft) below.

79.
0370. The minimum flight visibility required for VFR flights above 10,000 feet MSL and more than 1,200 feet AGL is

A— 1 SM.
B— 3 SM.
C— 5 SM.
D— not specified by regulation.

Answer (C) is correct (0370). *(FAR 91.105)*
Above 10,000 ft MSL and above 1,200 ft AGL, the visibility requirement is 5 statute miles.

Chapter 8: Federal Aviation Regulations

80.
0373. For VFR flight operations above 10,000 feet MSL and more than 1,200 feet AGL, the minimum horizontal distance from clouds required is

A— 1,000 feet.
B— 2,000 feet.
C— 1 mile.
D— 5 miles.

Answer (C) is correct (0373). *(FAR 91.105)*
In VFR above 10,000 ft MSL and above 1,200 ft AGL, the minimum horizontal distance from clouds is 1 statute mile.

81.
0375. The minimum distance from clouds required for VFR operations below 10,000 feet is

A— remain clear of clouds.
B— 500 feet below, 1,000 feet above, and 2,000 feet horizontally.
C— 500 feet above, 1,000 feet below, and 2,000 feet horizontally.
D— 1,000 feet above, 1,000 feet below, and 1 mile horizontally.

Answer (B) is correct (0375). *(FAR 91.105)*
Below 10,000 ft MSL and above 1,200 ft AGL, the required clearance from clouds is 500 ft below, 1,000 ft above, and 2,000 ft horizontally. This is true whether in or out of controlled airspace.

91.109 VFR CRUISING ALTITUDE OR FLIGHT LEVEL

82.
0383. Which VFR cruising altitude is acceptable for a flight on a magnetic course of 175°? The terrain is less than 1,000 feet.

A— 4,000 feet.
B— 4,500 feet.
C— 5,000 feet.
D— 5,500 feet.

Answer (D) is correct (0383). *(FAR 91.109)*
When operating above 3,000 ft AGL but less than 18,000 ft MSL on a magnetic course of 0° to 179°, fly at an odd thousand MSL altitude plus 500 ft. When on a magnetic course of 180° to 359°, fly at an even thousand foot altitude plus 500 ft. Thus, one should always be at a 1,000-ft altitude plus 500 ft, never at a multiple of a thousand feet.

When going toward the east (0° to 179°), do not fly at even; "E" in east does not stand for "E" in even. Thus, when flying a 175° magnetic course, one is flying toward the east and uses an odd thousand plus 500 ft altitude, e.g., 5,500 ft.

83.
0385. Each person operating an aircraft under VFR in level cruising flight at an altitude of more than 3,000 feet above the surface, and below 18,000 feet MSL, shall maintain an odd-thousand plus 500-foot altitude while on a

A— magnetic heading of 180° through 359°.
B— magnetic course of 0° through 179°.
C— true course of 180° through 359°.
D— true heading of 0° through 179°.

Answer (B) is correct (0385). *(FAR 91.109)*
When operating above 3,000 ft AGL but less than 18,000 ft MSL on a magnetic course of 0° to 179°, fly at an odd thousand MSL altitude plus 500 ft. When going toward the east (0° to 179°), do not fly at even; "E" in east does not stand for "E" in even.

84.
0384. Which VFR cruising altitude is acceptable for a flight on a magnetic course of 185°? The terrain is less than 1,000 feet.

A— 4,000 feet.
B— 4,500 feet.
C— 5,000 feet.
D— 5,500 feet.

Answer (B) is correct (0384). *(FAR 91.109)*
When on a magnetic course of 180° to 359°, fly at an even thousand foot altitude plus 500 ft. Thus, when flying a 185° magnetic course, one is flying toward the west and an even thousand plus 500 ft altitude is used, e.g., 4,500 ft.

91.163 MAINTENANCE, PREVENTIVE MAINTENANCE, AND ALTERATIONS

85.
0386. The responsibility for ensuring that an aircraft is maintained in an airworthy condition is primarily that of the

A— pilot in command of the aircraft.
B— owner or operator of the aircraft.
C— maintenance shop.
D— certified mechanic who signs the aircraft maintenance records.

Answer (B) is correct (0386). *(FAR 91.163)*
The owner or operator of an aircraft is primarily responsible for maintaining that aircraft in an airworthy condition.

91.167 OPERATION AFTER MAINTENANCE, PREVENTIVE MAINTENANCE, REBUILDING, OR ALTERATION

86.
0388. If an alteration or repair substantially affects an aircraft's operation in flight, that aircraft must be test flown by an appropriately-rated pilot and approved for return to service prior to being operated

A— by any private pilot.
B— with a passenger aboard.
C— for compensation or hire.
D— away from the vicinity of the airport.

Answer (B) is correct (0388). *(FAR 91.167)*
If an alteration or repair has been made that may have changed an airplane's flight characteristics, the airplane must be test flown and approved for return to service by an appropriately rated pilot prior to being operated with a passenger aboard. The test pilot must be at least a private pilot and appropriately rated for the airplane being tested.

87.
0389. Before a passenger can be carried in an aircraft that has been altered in a manner that may have appreciably changed its flight characteristics, it must be flight tested by an appropriately-rated pilot holding at least a

A— Commercial Pilot Certificate and an instrument rating.
B— Private Pilot Certificate.
C— Commercial Pilot Certificate and a mechanic's certificate.
D— Commercial Pilot Certificate.

Answer (B) is correct (0389). *(FAR 91.167)*
If an alteration or repair has been made that may have changed an airplane's flight characteristics, the airplane must be test flown and approved for return to service by an appropriately rated pilot prior to being operated with a passenger aboard. The test pilot must be at least a private pilot and appropriately rated for the airplane being tested.

91.169 INSPECTIONS

88.
0391. An aircraft's last annual inspection was performed on July 12, this year. The next annual inspection will be due no later than

A— July 13, next year.
B— July 31, next year.
C— 100 flight hours following the last annual inspection.
D— 12 calendar months after the date shown on the Airworthiness Certificate.

Answer (B) is correct (0391). *(FAR 91.169)*
Annual inspections expire on the last day of the 12th calendar month after the previous annual inspection. If an annual inspection is performed on July 12 of this year, it will expire at midnight on July 31 next year.

Answer (A) is incorrect because annual inspections always expire on the last day of the month of issuance. Answer (C) is incorrect because 100-hr inspections are required only for airplanes used for commercial purposes. Answer (D) is incorrect because the airworthiness certificate is only issued once, when the airplane is manufactured. Its date has no effect on the annual inspection.

91.173 MAINTENANCE RECORDS

89.
0390. Completion of an annual inspection and the return of the aircraft to service should always be indicated by

A— the relicensing date on the Registration Certificate.
B— an appropriate notation in the aircraft maintenance records.
C— an inspection sticker placed on the instrument panel that lists the annual inspection completion date.
D— the issuance date of the Airworthiness Certificate.

Answer (B) is correct (0390). *(FAR 91.173)*
 Completion of an annual airplane inspection must be evidenced by appropriate notations in the airplane maintenance records or logbooks.
 Answer (A) is incorrect because the registration certificate shows ownership. Answer (C) is incorrect because maintenance information is found in the airplane logbooks, not on inspection stickers. Answer (D) is incorrect because the airworthiness certificate is issued when the airplane is initially manufactured. It is irrelevant to the annual inspection.

90.
0392. To determine the expiration date of the last annual aircraft inspection, a person should refer to the

A— Airworthiness Certificate.
B— Registration Certificate.
C— aircraft maintenance records.
D— owner/operator manual.

Answer (C) is correct (0392). *(FAR 91.173)*
 After maintenance inspections have been completed, maintenance personnel should make the appropriate entries in the aircraft maintenance records or logbooks. This is where the date of the last annual inspection can be found.
 Answers (A), (B), and (D) are incorrect because each does not include maintenance information.

91.
0394. Which record or documents shall the owner or operator of an aircraft keep to show compliance with an applicable Airworthiness Directive?

A— The aircraft maintenance records.
B— Airworthiness Certificate and owner's handbook.
C— Airworthiness and Registration Certificates.
D— Aircraft flight manual and owner's handbook.

Answer (A) is correct (0394). *(FAR 91.173)*
 Maintenance records must include the current status of applicable Airworthiness Directives (ADs), including the method of compliance.
 Answers (B), (C), and (D) are incorrect because each contains no maintenance records or data on compliance with ADs.

NTSB 830 NATIONAL TRANSPORTATION SAFETY BOARD (Part 830)

92.
0305. If an aircraft is involved in an accident which resulted in substantial damage to the aircraft, the nearest NTSB field office should be notified

A— immediately.
B— within 5 days.
C— within 7 days.
D— within 10 days.

Answer (A) is correct (0305). *(NTSB 830.5)*
 An aircraft accident resulting in substantial damage must be reported immediately to the National Transportation Safety Board. A written report must be filed within 10 days of the accident.

93.
0310. The operator of an aircraft that has been involved in an accident is required to file an accident report within how many days?

A— 3.
B— 5.
C— 7.
D— 10.

Answer (D) is correct (0310). *(NTSB 830.15)*
 After the NTSB has been notified immediately of an airplane accident, a written accident report is also required to be filed within the next 10 days.

Chapter 8: Federal Aviation Regulations

94.
0309. An aircraft is involved in an accident that results in substantial damage to the aircraft, but no injuries to the occupants. When must the pilot or operator of the aircraft notify the nearest NTSB field office of the occurrence?

A— Immediately.
B— Within 48 hours.
C— Within 1 week.
D— Within 10 days.

Answer (A) is correct (0309). *(NTSB 830.5)*
An aircraft accident resulting in substantial damage must be reported immediately to the National Transportation Safety Board, regardless of injury to occupants. A written report must be filed within 10 days of the accident.

95.
0308. Which incident requires an immediate notification to the nearest NTSB field office?

A— A forced landing due to engine failure.
B— Landing gear damage, due to a hard landing.
C— Inability of any required crewmember to perform normal flight duties due to in-flight injury or illness.
D— Substantial aircraft ground fire with no intention of flight.

Answer (C) is correct (0308). *(NTSB 830.5)*
Regulations require immediate notification to the National Transportation Safety Board of the inability of required flight crewmembers to perform normal flight duties as a result of injury or illness.

96.
0306. Which incident would necessitate an immediate notification to the nearest NTSB field office?

A— Generator/alternator failure.
B— An in-flight fire.
C— Loss of VOR receiver capability.
D— Ground damage to the propeller blades.

Answer (B) is correct (0306). *(NTSB 830.5)*
Regulations require immediate notification to the NTSB of an aircraft accident resulting in substantial damage, serious injury, or death, or if any of the six incidents listed in 830.5 occur. An in-flight fire is one such incident.

Answers (A), (C), and (D) are incorrect because each does not constitute substantial damage under Part 830.5.

97.
0307. Which incident would require that an immediate notification be made to the nearest NTSB field office?

A— An overdue aircraft that is believed to be involved in an accident.
B— Radio (communication) failure.
C— Generator or alternator failure.
D— Near midair collision.

Answer (A) is correct (0307). *(NTSB 830.5)*
An overdue aircraft that is believed to be involved in an accident must be reported immediately to the National Transportation Safety Board.

98.
0311. The operator of an aircraft that has been involved in an incident is required to submit a report to the nearest field office of the NTSB

A— within 3 days.
B— within 7 days.
C— within 10 days.
D— only if requested to do so.

Answer (D) is correct (0311). *(NTSB 830.15)*
An incident is an occurrence other than an accident associated with the operation of an aircraft that affects or could affect the safety of operations. Such an incident (other than an accident) must be reported if requested by the NTSB.

171

CHAPTER NINE
NAVIGATIONAL PUBLICATIONS
AND SECTIONAL CHARTS

FAA Advisory Circulars	(4 questions)	171, 175
Airport/Facility Directory	(2 questions)	171, 177
Miscellaneous AIM Questions	(4 questions)	172, 177
Sectional Charts	(24 questions)	172, 178

This chapter contains outlines of major concepts tested; all FAA test questions and answers regarding *Airport/Facility Directories* (AFDs), *FAA Advisory Circulars*, and other navigational publications including sectional charts; and an explanation of each answer. The subtopics or modules within this chapter are listed above, followed in parentheses by the number of questions from the FAA written test pertaining to that particular module. The two numbers following the parentheses are the page numbers on which the outline and questions begin for that module.

The questions on sectional charts refer to Figure 23 and Figure 24. Figure 23 is on the inside of the back cover of this book and Figure 24 is on the outside of the back cover. The inside of the front cover contains the legend of a sectional chart. Remember in the *FAA Test Question Book* all figures and charts appear in the appendices at the back of the book.

CAUTION: Recall that the sole purpose of this book is to expedite your passing the FAA written test for the private pilot certificate. Accordingly, all extraneous material (i.e., topics or regulations not directly tested on the FAA written test) is omitted, even though much more information and knowledge are necessary to fly safely. This additional material is presented in *RECREATIONAL PILOT FLIGHT MANEUVERS* and *PRIVATE PILOT HANDBOOK*, available from Aviation Publications, Inc. See pages 208-210 for more information and an order form.

FAA ADVISORY CIRCULARS (4 questions)

1. The FAA issues advisory circulars to provide a systematic means for the issuance of nonregulatory material of interest to the aviation public.

    a. They may be ordered directly from the FAA.

2. The circulars are issued in a numbered system of general subject matter areas to correspond with the subject areas in Federal Aviation Regulations (FARs).

    a. **Airmen** has the subject number **60**.
    b. **Airspace** has the subject number **70**.
    c. **ATC and general operations** has the subject number **90**.

AIRPORT/FACILITY DIRECTORY (2 questions)

1. *Airport/Facility Directories* (AFDs) are published by the U.S. Department of Commerce every 2 months for each of seven geographical districts of the United States.

    a. *AFDs* provide information on services available, runways, special conditions at the airport, communications, navigation aids, etc.

172   Chapter 9: Navigational Publications and Sectional Charts

2. The third item on the first line is the number of miles and direction of the airport from the city.
   a. EXAMPLE. 1 NW means 1 mile northwest of the city.

3. Right-turn traffic is indicated by *Rgt tfc* following a runway number (remember all airport traffic patterns are with left turns unless otherwise indicated).

4. An *AFD* excerpt appears as Figure 29 on page 176.

5. The following 23 items are included in the *AFD* for each airport to the extent they exist at that airport.

1.	City/Airport Name	13.	Oxygen
2.	NOTAM Service	14.	Traffic Pattern Altitude
3.	Location Identifier	15.	Airport of Entry and Landing Rights Airports
4.	Airport Location	16.	Certificated Airport (FAR 139)
5.	Time Conversion	17.	FAA Inspection
6.	Geographic Position of Airport	18.	Runway Data
7.	Charts	19.	Airport Remarks
8.	Instrument Approach Procedures	20.	Weather Data Sources
9.	Elevation	21.	Communications
10.	Rotating Light Beacon	22.	Radio Aids to Navigation
11.	Servicing	23.	COMM/NAVAID Remarks
12.	Fuel		

MISCELLANEOUS AIM QUESTIONS (4 questions)

1. **AIM** is the *Airman's Information Manual* published by the FAA every 4 months and is the official FAA guide to basic flight information and ATC procedures.

2. **Military operations areas** (MOA) consist of airspace established for separating military training activities from IFR traffic.
   a. VFR traffic should exercise extreme caution when flying within an MOA.
   b. Information regarding MOA activity can be obtained from flight service stations (FSSs) within 100 miles of the activity.

3. **Alert areas** may contain a high volume of pilot training or other unusual activity.
   a. All pilots, both those using the area and those crossing it, are equally responsible for collision avoidance.

4. The minimum altitude to fly over a **wildlife refuge** is 2,000 ft AGL.

SECTIONAL CHARTS (24 questions)

1. Sectional charts are divided by horizontal lines of latitude and vertical lines of longitude, both of which are incremented by short tick marks along the entire line. Each tick represents one minute (1/60 of a degree).
   a. The lines of **longitude** and **latitude** on the sectional chart are 30 minutes of arc (½ of 1°) apart and are numbered "degrees and minutes," e.g., 31° or 31°30'.
   b. Longitude increases to the left (west) and latitude increases up (north).
   c. Each rectangular area bounded by lines of latitude and longitude contains a pair of numbers in large, bold print to indicate the height of the maximum elevation of terrain or obstructions within that area of latitude and longitude.
      1) The larger number to the left is thousands of feet.
      2) The smaller number to the right is hundreds of feet.

Chapter 9: Navigational Publications and Sectional Charts   173

2. Airspace is either controlled or uncontrolled in terms of VFR visibility and distance from cloud requirements.

   a. **Lower limits of controlled airspace are**
      1) Surface in control zones around airports marked by dashed lines,
      2) 700 ft AGL in transition areas marked by magenta (red) bands,
      3) 1,200 ft AGL in federal airways and other areas marked by blue bands.

   b. **Upper limits** are 14,500 ft MSL (base of the Continental Control Area).

ALTITUDE	UNCONTROLLED AIRSPACE		CONTROLLED AIRSPACE	
	Flight Visibility	Distance From Clouds	**Flight Visibility	**Distance From Clouds
1200'or less above the surface, regardless of MSL Altitude.	*1 statute mile	Clear of clouds	3 statute miles	500' below 1000' above 2000' horizontal
More than 1200'above the surface, but less than 10,000'MSL.	1 statute mile	500' below 1000' above 2000' horizontal	3 statute miles	500' below 1000' above 2000' horizontal
More than 1200'above the surface and at or above 10,000'MSL.	5 statute miles	1000' below 1000' above 1 statute mile horizontal	5 statute miles	1000' below 1000' above 1 statute mile horizontal

3. To determine the **magnetic course (MC) from one airport to another,** correct the true course (TC) only for magnetic variation, i.e., make no allowance for wind correction angle.

   a. Determine the true course by placing the straightedge of a navigational plotter or protractor along the route, with the hole in the plotter on the intersection of the route and a meridian (the vertical line with little crosslines).

      1) The true course is measured by the numbers on the protractor portion of the plotter (semi-circle) at the meridian.

      2) Note that up to four numbers (90° apart) are provided on the plotter. You must determine which is the direction of the flight.

   b. Alternatively, you can use a line of latitude (horizontal line with little crosslines) if your course is in a north or south direction.

      1) This is why there are four numbers on the plotter. You may be using either a meridian or line of latitude to measure your course and be going in either direction along the course line.

   c. Determine the **magnetic course** by adjusting the true course for magnetic variation (angle between true north and magnetic north).

      1) On the sectional charts, a long dashed line is provided with the number of degrees of magnetic variation. The variation is either east or west and is signified by "E" or "W," e.g., 3°E or 5°W.

      2) If the variation is east, subtract; if west, add. (Memory aid: east is least and west is best). This is from TC to MC.

4. The vertical limits of **Airport Radar Service Areas (ARSAs)** are marked with heavy dashed magenta (red) lines on aeronautical charts. The top limit is shown over a straight line and the bottom limit beneath the line.

   a. EXAMPLE. See Figure 24 on the outside back cover. At the lower right is the Savannah ARSA section. In the center appears 41 over a horizontal line over SFC.

1) $\frac{41}{SFC}$ means the top of the ARSA inner circle is 4,100 ft MSL and the bottom is the surface.

5. **Military flight routes** are charted with 4-digit numbers for those below 1,500 ft AGL.
   a. Those above 1,500 ft AGL are identified by 3-digit numbers.
   b. IR means the flights are made in accordance with IFR.
   c. VR would mean they use VFR.

6. **Alert areas** may contain a high volume of pilot training or other unusual activity. Pilots using the area as well as pilots crossing the area are equally responsible for collision avoidance.

7. **Airport identifiers** include the following information:
   a. The name of the airport.
   b. The elevation of the airport, followed by the length of the longest hard-surfaced runway. An L between the altitude and length indicates lighting.
      1) EXAMPLE: 1008 L 70 means 1,008 ft above MSL, L is for lighting sunset to sunrise, and the length of the longest hard-surfaced runway is 7,000 ft.
      2) If the L is with asterisk, it means pilot controlled lighting, e.g., by clicking one's microphone on a certain frequency.
   c. The UNICOM frequency if one has been assigned (e.g., 122.8) is shown after or underneath the runway length.
   d. At controlled airports, the tower frequency is usually under the airport name and above the runway information. The frequency is preceded by CT.
      1) If not a federal control tower, a NFCT precedes the CT frequency.

8. **Obstructions on sectional charts.**
   a. Obstructions of a height less than 1,000 ft AGL have the symbol ∧.
      1) A group of such obstructions has the symbol ⋀.
   b. Obstructions of a height of 1,000 ft or more AGL have the symbol ⊼.
      1) A group of such obstructions has the symbol ⋀.
   c. Obstructions with high intensity lights have arrows projecting from the top of the obstruction symbol.
   d. The actual height of the top of obstructions is listed near the obstruction by two numbers: one in bold print over another in light print with parentheses around it.
      1) The bold number is the elevation of the top of the obstruction in ft above MSL.
      2) The light number in parentheses is the height of the obstruction in ft AGL.
      3) The elevation (MSL) at the base of the obstruction is the bold figures minus the light figures.
         a) Use this computation to compute terrain elevation.
         b) Terrain elevation is also given in the airport identifier for each airport.
   e. ⊙ means a lookout tower.

9. Airports attended during normal business hours and having fuel service are indicated on airport symbols by the presence of small solid squares at the top and bottom and on both sides (9 o'clock and 3 o'clock) on the airport symbol.

Chapter 9: Navigational Publications and Sectional Charts

> **QUESTIONS AND ANSWER EXPLANATIONS**
>
> All of the FAA questions from the written test for the Recreational Pilot certificate relating to AFDs, FAA Advisory Circulars, and other navigational publications including sectional charts as outlined previously are reproduced below in the same modules as the previous outlines. To the immediate right of each question are the correct answer and answer explanation. You should cover these answers and answer explanations while responding to the questions. Refer to the general discussion in Chapter 1 on how to take the examination.
>
> Remember that the questions from the FAA's Written Test Book have been reordered by topic, and the topics have been organized into a meaningful sequence. Accordingly, the first line of the answer explanation gives the FAA question number and the citation of the authoritative source for the answer.

FAA ADVISORY CIRCULARS

1.
0312. FAA advisory circulars (some free, others at cost) are available to all pilots and are obtained by

A— distribution from the nearest FAA district office.
B— ordering those desired.
C— subscribing to the Federal Register.
D— subscribing to FAR's.

Answer (B) is correct (0312). *(AC 00-2)*
FAA Advisory Circulars are issued with the purpose of informing the public of nonregulatory material of interest. Advisory circulars are available by ordering directly from the FAA. You should write to: U.S. Department of Transportation, Public Agent Section N-443.1, Washington, D.C. 20590, for a copy of Advisory Circular 00-2, which provides complete information on advisory circulars. Some circulars are free. Those for which you must pay are available from the Superintendent of Documents, U.S. Government Printing Office, Washington, D.C. 20402.

2.
0495. FAA advisory circulars containing matter covering the subject "Airmen" are issued under which subject number?

A— 20.
B— 60.
C— 70.
D— 90.

Answer (B) is correct (0495). *(AC 00-2)*
The requirement concerns the Advisory Circular Subject Number for "airmen." In the FARs, airmen have the 60 series, e.g., Part 61 regards the certification of airmen.
Answer (A) is incorrect because subject number 20 deals with aircraft. Answer (C) is incorrect because subject number 70 deals with airspace. Answer (D) is incorrect because subject number 90 deals with ATC and general operations.

3.
0496. FAA advisory circulars containing subject matter specifically related to "Airspace" are issued under which subject number?

A— 70.
B— 90.
C— 120.
D— 150.

Answer (A) is correct (0496). *(AC 00-2)*
FAA Advisory Circulars regarding airspace have the subject number 70, just as the FARs regarding airspace have Part 70 numbers.
Answer (B) is incorrect because subject number 90 refers to ATC and general operations. Answer (C) is incorrect because subject number 120 refers to air carriers. Answer (D) is incorrect because subject number 150 refers to airport noise compatibility planning.

4.
0497. FAA advisory circulars containing subject matter specifically related to "Air Traffic Control and General Operations" are issued under which subject number?

A— 20.
B— 60.
C— 70.
D— 90.

Answer (D) is correct (0497). *(AC 00-2)*
FAA Advisory Circulars relating to air traffic control and general operations have the subject number 90, as do FARs on the same topics.
Answer (A) is incorrect because subject number 20 refers to aircraft. Answer (B) is incorrect because subject number 60 refers to airmen. Answer (C) is incorrect because subject number 70 refers to airspace.

NEBRASKA

LOUP CITY MUNI (NEØ3) 1 NW UTC-6(-5DT) 41°17'25"N 98°59'25"W OMAHA
  2070 B FUEL 100LL L-11B
  RWY 15-33: H3200X50 (ASPH) S-8 LIRL
    RWY 33: Trees.
  RWY 04-22: 2100X100 (TURF)
    RWY 04: Tree.   RWY 22: Road.
  AIRPORT REMARKS: Unattended. For svc call 308-745-0328.
  COMMUNICATIONS: CTAF 122.9
    COLUMBUS FSS (OLU) TF 1-800-WX-BRIEF. NOTAM FILE OLU.
  RADIO AIDS TO NAVIGATION: NOTAM FILE OLU.
    WOLBACH (H) VORTAC 114.8 OBH Chan 95 41°22'33"N 98°21'12"W 250° 29.3 NM to fld. 2010/10E.

MARTIN FLD (See SO SIOUX CITY)

§ **McCOOK MUNI** (MCK) 2 E UTC-6(-5DT) 40°12'23"N 100°35'29"W OMAHA
  2579 B S4 FUEL 100LL, JET A ARFF Index A H-2J, L-11A
  RWY 12-30: H5998X100 (CONC) S-30, D-38 MIRL .6% up NW IAP
    RWY 12: MALS. VASI(V4L)—GA 3.0°TCH 33'. Tree.   RWY 30: REIL. VASI(V4L)—GA 3.0°TCH 42'.
  RWY 03-21: H3999X75 (CONC) S-30, D-38 MIRL
    RWY 03: VASI(V2L)—GA 3.0°TCH 26'. Rgt tfc.   RWY 21: VASI(V2L)—GA 3.0°TCH 26'.
  RWY 17-35: 1350X200 (TURF)
  AIRPORT REMARKS: Attended daylight hours. ACTIVATE MALS Rwy 12—122.8. For VASI Rwy 12-30 Key 122.8 3 times.
    Closed to air carrier operations of aircraft with seating capacity over 30 passengers except with prior approval, call
    arpt manager 308-345-2022; 24 hours in advance. Control Zone effective 1100-0500Z‡. Except Holidays.
  COMMUNICATIONS: CTAF/UNICOM 122.8
    COLUMBUS FSS (OLU) TF 1-800-WX-BRIEF. NOTAM FILE MCK
    RCO 122.6 (COLUMBUS FSS)
  ® DENVER CENTER APP/DEP CON 132.7
  RADIO AIDS TO NAVIGATION: NOTAM FILE MCK.
    (L) VORW/DME 110.0 MCK Chan 37 40°12'14"N 100°35'38"W at fld. 2550/11E.

MEAD

UNIV NEB FIELD LAB (NEØ7) 3 SE UTC-6(-5DT) 41°10'45"N 96°27'40"W OMAHA
  1180
  RWY 17-35: 3000X100 (TURF)
    RWY 35: Road.
  AIRPORT REMARKS: Unattended.
  COMMUNICATIONS: CTAF 122.9
    COLUMBUS FSS (OLU) TF 1-800-WX-BRIEF. NOTAM FILE OLU.

MILLARD (See OMAHA)

MILLER FLD (See VALENTINE)

MINDEN

PIONEER VILLAGE FLD (ØV3) 1 NE UTC-6(-5DT) 40°30'47"N 98°56'42"W OMAHA
  2159 B S4 FUEL 100LL L-11B
  RWY 16-34: H3900X60 (CONC) S-28, D-52 MIRL
    RWY 16: PAPI(P2L)—GA 3.0° TCH 40'. Crops.   RWY 34: PAPI(P2L)—GA 3.0° TCH 40'. Building. Rgt tfc.
  RWY 05-23: 2175X300 (TURF)
    RWY 05: Trees.   RWY 23: Fence. Rgt tfc.
  AIRPORT REMARKS: Attended 1400-2300Z‡. For service after hours call 308-832-2809 or 832-2772. Tower 1163'
    AGL, 3240' MSL 9 NM ENE.
  COMMUNICATIONS: CTAF/UNICOM 122.8
    COLUMBUS FSS (OLU) TF 1-800-WX-BRIEF. NOTAM FILE OLU.
  RADIO AIDS TO NAVIGATION: NOTAM FILE HSI.
    HASTING (L) VOR/DME 108.8 HSI Chan 25 40°36'16"N 98°25'45"W 247° 24.2 NM to fld. 1950/10E.

FIGURE 29.—Airport/Facility Directory Excerpt.

AIRPORT/FACILITY DIRECTORY

5.
0281. (Refer to figure 29.) Traffic patterns in effect at McCook Municipal are

A— left-hand on all runways.
B— left-hand on Rwy 12, Rwy 21, and Rwy 30; right-hand on Rwy 03.
C— right-hand on all runways.
D— right-hand on the concrete runways and left-hand on the turf runways.

Answer (B) is correct (0281). *(AFD)*
McCook Municipal Airport's traffic pattern is right on runway 3, as indicated by "Rgt tfc" in the line beginning RWY 03.
Answers (A), (C), and (D) are incorrect because the traffic pattern is left turns unless the AFD indicates "Rgt tfc" in the runway information. Here, only runway 3 is marked as right turns.

6.
0282. (Refer to figure 29.) Where is Loup City Municipal located with relation to the city?

A— Northwest approximately 1 NM.
B— Northeast approximately 3 NM.
C— East approximately 10 NM.
D— 250°, 29.3 NM.

Answer (A) is correct (0282). *(AFD)*
Loup City Municipal Airport is located 1 mile northwest of the city, as indicated by "1 NW" on the first line of the directory for the airport.
Answer (B) is incorrect because "(NE03)" is the airport identifier. Answers (C) and (D) are incorrect because they refer to the location of the Wolbach VOR.

MISCELLANEOUS AIM QUESTIONS

7.
0273. When operating VFR in a MOA, a pilot

A— must obtain a clearance from the controlling agency prior to entering the MOA.
B— may operate only on the airways that transverse the MOA.
C— should exercise extreme caution when military activity is being conducted.
D— must operate only when military activity is not being conducted.

Answer (C) is correct (0273). *(AIM Para 114)*
Military operations areas consist of airspace established for separating military training activities from IFR traffic. VFR traffic should exercise extreme caution when flying within an MOA. Information regarding MOA activity can be obtained from flight service stations (FSSs) within 100 miles of the activity.
Answer (A) is incorrect because a clearance is not required to enter an MOA. Answer (B) is incorrect because authorized VFR flights may fly anywhere in the MOA. Answer (D) is incorrect because VFR flights are allowed when military activity is conducted, but pilots should exercise extreme caution.

8.
0210. Pilots desiring to know the current activity status of a Military Operations Area (MOA) should

A— refer to the most recently published sectional chart depicting the MOA.
B— refer to the Airman's Information Manual.
C— refer to the remarks section of the MOA manual.
D— contact an appropriate FSS.

Answer (D) is correct (0210). *(AIM Para 114)*
MOAs consist of airspace established for the purpose of separating certain military training activities from IFR traffic. The training activities necessitate aerobatic or abrupt flight maneuvers. Activity status of MOAs changes frequently and pilots can contact any Flight Service Station within 100 miles of the area to obtain real time information concerning the MOA's hours of operation. Also, ATC frequencies will be made available so that you can contact the controlling agency for traffic advisories.
Answer (A) is incorrect because sectional charts show MOAs but are not real-time in providing current activity. Answer (B) is incorrect because *AIM* has only a general description of MOAs. Answer (C) is incorrect because *MOA Manual* is a nonsense concept.

178 Chapter 9: Navigational Publications and Sectional Charts

9.
0274. Who is responsible for collision avoidance in an alert area?

A— The controlling agency.
B— All pilots, without exception.
C— Only the pilots transitioning the area.
D— All pilots except those participating in the training operations.

Answer (B) is correct (0274). *(AIM Para 115)*
Alert areas may contain a high volume of pilot training or other unusual activity. Pilots using the area as well as pilots crossing the area are equally responsible for collision avoidance.
Answer (A) is incorrect because pilots, not controlling agencies, are responsible. Answers (C) and (D) are incorrect because both participating and transiting pilots are responsible.

10.
0275. What minimum altitude is requested for aircraft over national wildlife refuges?

A— 500 feet AGL.
B— 1,000 feet AGL.
C— 2,000 feet AGL.
D— 3,000 feet AGL.

Answer (C) is correct (0275). *(AIM Para 563)*
The Fish and Wildlife Service requests that pilots maintain a minimum altitude of 2,000 ft above the terrain of wildlife refuge areas.

SECTIONAL CHARTS

Refer to Figure 23 inside the back cover for the following questions.

11.
0197. (Refer to figure 23, area B.) Under what conditions may a recreational pilot operate at Coeur D'Alene Airport?

A— When the ceiling is more than 1,000 feet and the visibility is more than 3 miles.
B— When the ceiling is more than 3,000 feet and the visibility is more than 1 mile.
C— When the tower is closed, the ceiling is at least 1,000 feet, and the visibility is at least 3 miles.
D— At any time.

Answer (C) is correct (0197). *(FAR 61.101)*
The Coeur D'Alene Airport, near B in Figure 23, has a control tower indicated by CT 119.1 in the airport information block to the upper right of the airport symbol. Since recreational pilots are not permitted to act as pilot-in-command in areas requiring communication with ATC, the only time a recreational pilot can operate at Coeur D'Alene Airport is when the tower is closed. Additionally, recreational pilots are precluded from flying when visibility is less than 3 miles. Additionally, the VFR weather minimums apply in the control zone and accordingly a ceiling of at least 1,000 ft is required.

12.
0198. (Refer to figure 23.) What is the magnetic course between St. Maries Airport (area D) and Shoshone County Airport (area C)?

A— 030°.
B— 068°.
C— 210°.
D— 248°.

Answer (A) is correct (0198). *(ACL)*
St. Maries Airport is at "D" and Shoshone County Airport is at "C." It is a northeasterly route and note that parallel to it is the magnetic variation dashed line of 19°30' East, which means you subtract 19°30' from the true course to obtain the magnetic course. Using a plotter, one determines that the true course is 50°. Subtracting 19°30' gives approximately a 30° magnetic course.
Answer (B) is incorrect because you subtract, not add, easterly magnetic variation. Answers (C) and (D) are incorrect because you are going northeast from St. Maries to Shoshone, not southwest.

Chapter 9: Navigational Publications and Sectional Charts

Refer to Figure 23 inside the back cover for the following questions.

13.
0199. (Refer to figure 23, area A.) What is the approximate latitude and longitude of County City Airport?

A— 48°18'N, 116°33'W.
B— 48°48'N, 116°33'W.
C— 48°48'N, 117°03'W.
D— 48°18'N, 117°03'W.

Answer (A) is correct (0199). *(ACL)*
County City Airport is at the top of the map. Note that longitude and latitude lines of the vertical and horizontal lines with the small tick marks on them. They are presented every 30 minutes which is ½ of 1°. About 2/3 of the way up on the very right side of Figure 23, there is a 48° which indicates the latitude of the County will be less than 48°30' because County City is below the 48°30' latitude line. Just below and to the left of the 48° is the 116° longitude line. The next line over is 116°30' and County City is about 3 minutes west of the 116°30' longitude line. Thus the latitude is 48°18' north and the longitude is 116°33' west.

14.
0200. (Refer to figure 23, area D.) Identify the airspace overlying St. Maries Airport from the surface to 14,500 feet MSL.

A— Uncontrolled airspace - surface to 14,500 feet.
B— Uncontrolled airspace - surface to 700 feet AGL, control area - 700 feet AGL to 14,500 feet MSL.
C— Uncontrolled airspace - surface to 14,500 feet MSL.
D— Control area - surface to 14,500 feet MSL.

Answer (C) is correct (0200). *(ACL)*
St. Maries Airport is just below "D" on Figure 23. Note that the controlled airspace is marked by blue shading. You can see the band of controlled airspace about victor airways on the map. At "D" you see St. Maries is just outside of controlled airspace and accordingly from the surface to 14,500 ft MSL (where the continental control area begins), the airspace is uncontrolled.

Answer (A) is incorrect because it does not indicate whether it is 14,500 ft AGL or MSL. Answer (B) is incorrect because it describes a situation within a magenta area such as the restricted airport just to the west of Henley which is at "B." Note it is in an area that is within a magenta shading and this is where controlled airspace begins at 700 ft AGL and goes to the base of the continental control area. Answer (D) is incorrect because controlled airspace from the surface up occurs inside of control zones. The only control zone on Figure 23 is that around the Coeur D'Alene Airport.

15.
0201. (Refer to figure 23.) What type of military flight operations should a pilot expect along IR340?

A— Area navigation training flights in IMC weather.
B— Helicopter instrument training flights below 1,200 feet AGL.
C— Airplane instrument training flights below 1,500 feet AGL at speeds in excess of 150 knots.
D— Airplane training flights above 1,500 feet AGL regardless of the weather and at speeds in excess of 250 knots.

Answer (D) is correct (0201). *(AIM Para 132)*
IR #340 is a military training route. It extends in gray from down near St. Maries at "D" up to Pend Oreille Lake and then to the northwest. The IR prefix indicates IFR rather than VFR which is indicated by V/R. The 3-digit identifier number indicates above 1,500 ft AGL versus a 4-digit identifier which would indicate at or below 1,500 ft.

Answer (A) is incorrect because they are specific routes, not an area. Answer (B) is incorrect because the routes for military aircraft are not restricted to helicopters and they are either above or below 1,500 ft, not 1,200 ft. Answer (C) is incorrect because if the flight is to be less than 1,500 ft AGL it will have a 4-digit identifier.

Refer to Figure 23 inside the back cover for the following questions.

16.
0202. (Refer to figure 23, area B.) What type of airspace is overlying Henley Airport?

A— Uncontrolled airspace from the surface to 14,500 feet MSL.
B— Uncontrolled airspace from the surface to 1,200 feet AGL, then control area from 1,200 feet AGL to 14,500 feet MSL.
C— Controlled airspace from the surface to 10,000 feet MSL.
D— Uncontrolled airspace from the surface to 700 feet AGL, then controlled airspace from 700 feet AGL to 14,500 feet MSL.

Answer (B) is correct (0202). *(ACL)*
Henley Airport is just above "B" on Figure 23. It is within a blue shaded area which means controlled airspace begins 1,200 ft AGL (i.e., uncontrolled airspace exists from the surface up to 1,200 ft AGL) and is controlled up to the base of the continental control area.

Answer (A) is incorrect because Henley Airport is not within a control zone and thus controlled airspace begins at 1,200 ft AGL, not the surface. Answer (C) is incorrect because airspace is controlled up to the continental control area (14,500 ft MSL). Answer (D) is incorrect because it describes a transition area such as immediately to the west and south of Henley Airport which is within a magenta shaded area.

17.
0203. (Refer to figure 23, area A.) What is the magnetic variation for a trip from Priest River Airport to CX Airport?

A— 020° east.
B— 020° west.
C— 097° east.
D— 277° west.

Answer (A) is correct (0203). *(ACL)*
Priest River Airport is in the upper left of Figure 23 and CX Airport is a restricted private field just on the east side of Pend Oreille Lake. The line of magnetic variation is a dashed magenta line just to the east of Priest River that runs from southwest to northeast of Pend Oreille Lake. Note that it indicates 20° east.

18.
0204. (Refer to figure 23.) Which public use airports are indicated as having fuel and are attended during normal working hours?

A— Coeur D'Alene, Magee, and County City.
B— Lake Pend Oreille, St. Maries, and Brooks.
C— Shoshone County, Henley, and Lake Pend Oreille.
D— Coeur D'Alene, Priest River, and Brooks.

Answer (C) is correct (0204). *(ACL)*
Public-use airports having fuel and attended during normal working hours have tiny square blocks on their round airport symbol, usually at the 12, 3, 6, and 9 o'clock positions. In order to answer this question, you need to find all of the airports listed. Y = Yes, fuel services are available.

	Airport	Location	Fuel
A	Coeur D'Alene	1" SW of B	Y
	Magee	2" right of B	N
	County City	1" above A	Y
B	Lake Pend Oreille	1" NE of A	Y
	St. Maries	Just below D	N
	Brooks	1½" below B	Y
C	Shoshone County	Just below C	Y
	Henley	Just above B	Y
	Lake Pend Oreille	1" NE of A	Y
D	Coeur D'Alene	SW of B	Y
	Priest River	Upper left corner	N
	Brooks	1½" below B	Y

Chapter 9: Navigational Publications and Sectional Charts

Refer to Figure 23 inside the back cover for the following questions.

19.
0205. (Refer to figure 23, area B.) What is the maximum elevation, in feet, bounded by the ticked lines of latitude and longitude?

A— 4,061 feet MSL.
B— 5,241 feet MSL.
C— 5,665 feet MSL.
D— 6,000 feet MSL.

Answer (D) is correct (0205). (ACL)
In Area "B," middle left, the maximum elevation within that box of latitude and longitude lines is 6,000 ft, indicated by the large 6 and the slightly smaller 0 to the immediate right of the City of Coeur D'Alene.

Answer (A) is incorrect because the 4,061 ft MSL is the AGL height of the tower just to the right of the Walla Walla Flight Watch Communications Box which is halfway between "D" and "B." Answer (B) is incorrect because 5,241 ft MSL is the height of Mica Peak which is to the southwest of the City of Coeur D'Alene right at the edge of Figure 23. Answer (C) is incorrect because 5,665 ft MSL is the elevation of a mountain just to the right of "B."

20.
0206. (Refer to figure 23, area A.) What is the height of the obstacle located approximately 3 NM southeast of CX Airport?

A— 240 feet MSL.
B— 2,400 feet AGL.
C— 2,563 feet AGL.
D— 2,563 feet MSL.

Answer (D) is correct (0206). (ACL)
CX Airport in Figure 23 is on the east side of Pend Oreille Lake. To the southeast of CX Airport is a tower with a an elevation of 2,563 ft MSL and 240 ft AGL.

Answer (A) is incorrect because the tower is 240 ft AGL not MSL. Answer (B) is incorrect because the 240 indicates 240 ft not 2,400 ft. Answer (C) is incorrect because the 2,563 ft is MSL, not AGL.

21.
0207. (Refer to figure 23.) What feature is designated by the symbol ⊙ ?

A— A mine or quarry.
B— An oil well.
C— A water tank.
D— A lookout tower.

Answer (D) is correct (0207). (ACL)
A circle with a solid triangle in the center of it is a lookout tower.

Answers (A), (B), and (C) are incorrect because their symbols are

● Tank-water, oil or gas
○ Oil Well
⚒ Mines And Quarries

Refer to Figure 24 on the outside of the back cover for the following questions.

22.
0208. (Refer to figure 24.) What is the upper limit of the Savannah ARSA?

A— Up to but not including 3,000 feet AGL.
B— The base of the Continental Control Area.
C— 4,000 feet above the primary airport.
D— 10,000 feet MSL.

Answer (C) is correct (0208). (ACL)
The upper limit of ARSAs is approximately 4,000 ft AGL above the primary airport. Savannah International is 51 ft MSL as indicated by the airport identifier which is approximately one inch to the right of the Savannah International Airport on Figure 24. Each sector of the Savannah ARSA (which is marked by the thick magenta broken lines) has a number on top of a horizontal line and either a number or the three letters SFC printed below it. The numbers on top of the line indicate the ceiling of the ARSA for that particular sector, and the number or letters below that line indicating the floor of the ARSA for that particular sector.

Answer (A) is incorrect because 3,000 ft AGL is the ceiling of an airport traffic area. Answer (B) is incorrect because the base of the continental control area is 14,500 MSL. Answer (D) is incorrect because 10,000 ft MSL is a level above and below which there are different VFR minimum visibility and horizontal distances from cloud restrictions.

182 Chapter 9: Navigational Publications and Sectional Charts

Refer to Figure 24 on the outside of the back cover for the following questions.

23.
0209. (Refer to figure 24, area C.) What is the height of the lighted obstacle approximately 8 miles southwest of Savannah International?

A— 1,500 feet AGL.
B— 1,531 feet AGL.
C— 1,532 feet AGL.
D— 1,549 feet AGL.

Answer (C) is correct (0209). *(ACL)*

Eight miles southwest of Savannah, there is a single tall obstruction with sparks emanating from its tip which indicates it is lighted. The height of the obstruction is 1,532 AGL as indicated in the lighter print in parens beneath the 1,549 which is the height above (or in) MSL.

Answer (A) is incorrect because 1,500 is the MSL height of the double obstruction to the left of the lighted single obstruction. Answer (B) is incorrect because 1,532, not 1,531, is the AGL height of the obstruction. Answer (D) is incorrect because 1,549 is the MSL, not AGL, height of the obstruction.

24.
0211. (Refer to figure 24, area A.) What is the approximate latitude and longitude of Allendale County Airport?

A— 32°29'N, 81°16'W.
B— 32°29'N, 81°46'W.
C— 32°59'N, 81°16'W.
D— 32°59'N, 81°46'W.

Answer (C) is correct (0211). *(ACL)*

Allendale Airport is one inch above "A" in Figure 24. It is almost right on a latitude line. Note the longitude lines at the bottom of Figure 24. Allendale is approximately halfway between 81° and 81°30' west. Note the line of latitude at the bottom of the Figure 24 is 32° and thus the latitude line through the middle of Figure 24 is 32°30' and the line of latitude at the top of the chart (through Allendale) is 33°. Thus the latitude is 32°59' N and the longitude 81°16' W.

25.
0212. (Refer to figure 24, area A.) What is the magnetic course from Allendale County Airport to Plantation Airport?

A— 045°.
B— 051°.
C— 215°.
D— 221°.

Answer (D) is correct (0212). *(ACL)*

Allendale County Airport is one inch above "A" and Plantation Airport is southwest of the Allendale Airport about 3" on Figure 24. Note that Plantation Airport is directly on Airway 157 which has a 216° magnetic course from Allendale VOR. Since the Allendale Airport is just south of the Allendale VOR, it would require a slightly more westerly (i.e., 221° magnetic course).

Answers (A) and (B) are incorrect because the course is southwest not northeast. Answer (C) is incorrect because the course is more westerly than 216°, not less westerly.

26.
0213. (Refer to figure 24, area B.) The airspace overlying and within 5 SM of Statesboro Airport is

A— an Airport Traffic Area from the surface up to and including 3,000 feet AGL.
B— a control zone from the surface to 14,500 feet.
C— special use airspace from the surface to the overlying control area.
D— uncontrolled airspace from the surface to 700 feet AGL.

Answer (D) is correct (0213). *(ACL)*

The Statesboro Airport is within a magenta shaded area which means controlled airspace begins at 700 ft AGL. Thus, there is uncontrolled airspace from the surface to 700 ft AGL.

Answer (A) is incorrect because there is no control tower at Statesboro as evidenced by the lack of a CT plus a frequency in the airport identifier. Answer (B) is incorrect because a control zone is marked by a dashed blue line. Answer (C) is incorrect because there is no marking or mention of special airspace.

Chapter 9: Navigational Publications and Sectional Charts   183

Refer to Figure 24 on the outside of the back cover for the following questions.

27.
0214. (Refer to figure 24.) What type of military flight operations should a pilot expect along VR1059?

A— Instrument training flights at or below 1,500 feet MSL at speeds in excess of 250 knots.
B— Visual training flights at or below 1,500 feet AGL at speeds in excess of 250 knots.
C— Instrument training flights above 1,500 feet MSL at speeds in excess of 200 knots.
D— Visual training flights at or above 1,500 feet MSL at speeds in excess of 250 knots.

Answer (B) is correct (0214). *(ACL)*
Military training route VR1059 extends from one inch from the top of Figure 24 on the left side to approximately 2" from the top of Figure 24 on the right side. The VR indicates a VFR training route. The 4-digit identifying number indicates it is at or below 1,500 ft.
Answers (A) and (C) are incorrect because the VR indicates VFR rather than IFR. Answers (C) and (D) are incorrect because the 4-digit number indicates that the flights are at or below 1,500 ft, not above. Answer (C) is also incorrect because military training routes also indicate speeds in excess of 250 kts.

28.
0215. (Refer to figure 24, area C.) The elevation of Savannah International Airport is

A— 51 feet MSL.
B— 90 feet MSL.
C— 5,100 feet MSL.
D— 9,000 feet MSL.

Answer (A) is correct (0215). *(ACL)*
Savannah International Airport (lower right of Figure 24) has an elevation of 51 ft MSL. Note the airport identifier, which is an inch and a half up from the bottom on the right side of Figure 24.
Answers (B) and (D) are incorrect because the 90 indicates the length in hundreds of feet of the longest runway. Answer (C) is incorrect because one does not add zeros to the elevation MSL digits.

29.
0216. (Refer to figure 24.) Which public use airports are indicated as having fuel and are attended during normal working hours?

A— Davis (north of Savannah) and Savannah International Airports.
B— Allendale and Claxton-Evans Airports.
C— Cartwright and Savannah International Airports.
D— Plantation and Hampton Varnville Airports.

Answer (B) is correct (0216). *(ACL)*
Public use airports having fuel and being attended are indicated by small squares extending from the airport symbol at the 12, 3, 6, and 9 o'clock positions. To answer this question, you need to check all of the airports listed in the answers.
For "A," Davis Airport is about 1½ inches below "A" and does not sell fuel. Allendale, which is just above "A" on Figure 24, offers fuel and Claxton-Evans, which is to the left of "B," also offers fuel, so Answer (B) is correct. For "C," Cartwright is an army airport about 1 inch southwest of "B" and does not offer fuel. For "D," Hampton-Varnville which is 1 inch to the right of "A" does not offer fuel.

30.
0217. (Refer to figure 24, area A.) What is the maximum elevation, in feet, bounded by the ticked lines of latitude and longitude?

A— 436 feet MSL.
B— 600 feet MSL.
C— 3,000 feet MSL.
D— 6,000 feet MSL.

Answer (B) is correct (0217). *(ACL)*
The sectional chart block surrounded by the longitude and latitude lines enclosing "A" has a maximum elevation of 600 ft indicated by the large 0 and the slightly smaller 6 which is about 1 inch below "A."
Answer (A) is incorrect because 436 is the height of the tower at Newington which is just below the Sylvania NDB identification box. Answer (C) is incorrect because the 300 in brackets just below the 440 all of which is to the immediate right of the large 0 and slightly smaller 6 is the AGL height (after you add a zero) of the radio antenna just north of Estil. Answer (D) is incorrect because one adds two zeros, not three zeros, to the 6 to make 600, not 6,000.

Refer to Figure 24 on the outside of the back cover for the following questions.

31.
0218. (Refer to figure 24, area C.) What is the height of the lighted stack located approximately 14 NM north of Savannah International Airport?

A— 400 feet MSL.
B— 430 feet MSL.
C— 4,000 feet AGL.
D— 4,300 feet AGL.

Answer (B) is correct (0218). *(ACL)*
Located 14 NM north of Savannah International Airport is a 430 over the word "stacks." Stacks obstruction indicates high-intensity lights due to the sparks emanating from the tip. The 430 means 430 ft MSL.
Answer (A) is incorrect because there are two 400-ft obstructions, one MSL but they are in the vicinity of Springfield which is to about 14 NM northwest rather than north of Savannah International. Answers (C) and (D) are incorrect because you do not add zeroes to obstruction heights.

32.
0219. (Refer to figure 24, area C.) What is the minimum altitude a pilot should maintain above the Savannah National Wildlife Refuge?

A— 500 feet from any vessel, vehicles, person, or structure.
B— 1,000 feet AGL.
C— 1,200 feet AGL.
D— 2,000 feet AGL.

Answer (D) is correct (0219). *(AIM 565)*
All aircraft are requested to maintain a minimum altitude of 2,000 ft AGL over national parks, monuments, seashores, lakeshores, recreational areas, and scenic riverways administered by the National Park Service. The same is true for National Wildlife Refuges, Big Game Refuges, Game Ranges, and Wildlife Ranges administered by the U.S. Fish and Wildlife Service and wilderness and primitive areas administered by the U.S. Forest Service.

33.
0220. (Refer to figure 24, area C.) When, if ever, may a recreational pilot act as pilot in command on a flight to Savannah International Airport?

A— Anytime the control tower is operating.
B— When the control tower is not operating, the ceiling is more than 1,000 feet AGL, and the ground or surface visibility is 2 SM or more.
C— Anytime the ceiling is 1,000 feet or greater, and the surface visibility is 1 SM or more.
D— Never; the airport is located in an ARSA.

Answer (D) is incorrect (0220). *(FAR 61.101)*
Recreational Pilots cannot act as pilot-in-command on a flight to the Savannah International Airport because it is in an ARSA which requires ATC communication.
Answer (A) is incorrect because recreational pilots are prohibited from entering airspace that requires communication with ATC. Answers (B) and (C) are incorrect because recreational pilots are not permitted to fly when surface visibility is less than 3 statute miles.

34.
0221. (Refer to figure 24, area B.) In what type airspace is Camp Oliver Army Airfield located?

A— MOA.
B— Controlled firing area.
C— Restricted area.
D— Prohibited area.

Answer (C) is correct (0221). *(ACL)*
Camp Oliver Army Air Field is located in the lower left corner of Figure 24. It is in restricted area 3005A. Camp Oliver is contiguous to Cartwright.
Answers (A), (B) and (D) are incorrect because the airspace is restricted, not a military operation, controlled firing, or prohibited area.

CHAPTER TEN
FLIGHT PHYSIOLOGY AND FLIGHT OPERATIONS

Hyperventilation	(3 questions)	185, 188
Spatial Disorientation	(2 questions)	185, 189
Carbon Monoxide	(2 questions)	186, 189
Preflight Inspection	(4 questions)	186, 190
Starting the Engine	(2 questions)	186, 191
Taxiing Technique	(7 questions)	186, 191
Airspeed	(5 questions)	187, 193
Collision Avoidance	(2 questions)	187, 194

This chapter contains outlines of major concepts tested, all FAA test questions and answers regarding flight physiology and flight operations, and an explanation of each answer. The subtopics or modules within this chapter are listed above, followed in parentheses by the number of questions from the FAA written test pertaining to that particular module. The two numbers following the parentheses are the page numbers on which the outline and questions begin for that module.

CAUTION: Recall that the sole purpose of this book is to expedite your passing the FAA written test for the recreational pilot certificate. Accordingly, all extraneous material (i.e., topics or regulations not directly tested on the FAA written test) is omitted, even though much more information and knowledge are necessary to fly safely. This additional material is presented in *RECREATIONAL PILOT FLIGHT MANEUVERS* and *PRIVATE PILOT HANDBOOK*, available from Aviation Publications, Inc. See pages 208 to 210 for more information and an order form.

HYPERVENTILATION (3 questions)

1. Hyperventilation occurs when an excessive amount of air is breathed out of the lungs, e.g., when one becomes excited, undergoes stress, tension, fear, or anxiety.

    a. This results in an excessive **amount of carbon dioxide passed out of the body** and too much oxygen retained.

    b. The symptoms are dizziness, hot and cold sensations, nausea, etc.

2. Overcome hyperventilation symptoms by slowing the breathing rate, breathing into a bag, or by talking aloud.

SPATIAL DISORIENTATION (2 questions)

1. Spatial disorientation is a state of temporary confusion resulting from misleading information being sent to the brain by various sensory organs. The condition is sometimes called **vertigo.**

2. If you lose outside visual references and become disoriented, you are experiencing spatial disorientation. This occurs when you rely on the sensations of muscles and inner ear to tell you what the airplane's attitude is.

CARBON MONOXIDE (2 questions)

1. Blurred (hazy) thinking, uneasiness, dizziness, and tightness across the forehead are early symptoms of carbon monoxide poisoning. They are followed by a headache and, with large accumulations of carbon monoxide, a **loss of muscle power**.

2. Increases in altitude increase susceptibility to carbon monoxide poisoning because of decreased oxygen availability.

PREFLIGHT INSPECTION (4 questions)

1. Prior to **every flight**, the airplane should be preflighted.
    a. If the airplane has been used very recently, i.e., the same day, a walk-around inspection of the airplane should be made at a minimum.
    b. If the airplane had been preflighted the evening before and hangared, it would still be necessary to check for fuel contamination from condensation.
    c. If an airplane has been stored for a long time, in addition to the usual procedures, the airplane should be checked for damage or obstructions caused by animals, birds, or insects.

2. In summary, a **written checklist** for preflight inspection ensures that all necessary items are checked in a logical sequence.

STARTING THE ENGINE (2 questions)

1. After the engine starts, the throttle should be adjusted for proper RPM and the engine gauges checked.

2. When starting an airplane engine by hand, it is extremely important that a **competent pilot** be at the controls **in the cockpit**.

TAXIING TECHNIQUE (7 questions)

1. When taxiing in strong **quartering headwinds**, the aileron should be up on the side from which the wind is blowing.
    a. The elevator should be in the neutral position for tricycle-geared aircraft.
    b. The elevator should be in the up position for tailwheel aircraft.

2. When taxiing during strong **quartering tailwinds**, the aileron should be down on the side from which the wind is blowing.
    a. The elevator should also be down (control forward).

3. When taxiing high-wing, nosewheel-equipped airplanes, the most critical wind condition is a quartering tailwind.

AIRSPEED (5 questions)

1. When turbulence is encountered, the airplane's airspeed should be reduced to **maneuvering speed**.

    a. The pilot should attempt to maintain a level flight attitude, which means keeping control of the airplane (wings level).

    b. Constant altitude and constant airspeed are usually impossible and result in additional control pressure, which adds stress to the airplane.

2. In the event of a power failure after becoming airborne, the most important thing to do is to **maintain safe airspeed**.

    a. Do not maintain altitude at the expense of airspeed or a stall/spin will result.

3. **Best rate-of-climb** airspeed (V_Y) results in the most altitude gain per unit of time.

    a. It is used on normal takeoffs until a safe altitude, e.g., 1,000 ft AGL is reached.

4. **Best angle-of-climb** airspeed (V_X) results in the most altitude gain per unit of distance.

    a. It is used for obstacle clearance.

COLLISION AVOIDANCE (2 questions)

1. Any aircraft that appears to have no relative motion with respect to your aircraft and stays in one scan quadrant is likely to be on a collision course.

    a. If it increases in size, you should take immediate evasive action.

2. Prior to each maneuver, a pilot should visually scan the entire area for collision avoidance.

    a. When climbing (descending) VFR on an airway, you should execute gentle banks left and right to facilitate scanning for other aircraft.

Chapter 10: Flight Physiology and Flight Operations

> **QUESTIONS AND ANSWER EXPLANATIONS**
>
> All of the FAA questions from the written test for the Recreational Pilot certificate relating to flight physiology and flight operations and the material outlined previously are reproduced below in the same modules as the previous outlines. To the immediate right of each question are the correct answer and answer explanation. You should cover these answers and answer explanations while responding to the questions. Refer to the general discussion in Chapter 1 on how to take the examination.
>
> Remember that the questions from the FAA's Written Test Book have been reordered by topic, and the topics have been organized into a meaningful sequence. Accordingly, the first line of the answer explanation gives the FAA question number and the citation of the authoritative source for the answer.

HYPERVENTILATION

1.
0488. Rapid or extra deep breathing can cause a condition known as

A— hypoxia.
B— aerosinusitis.
C— aerotitis.
D— hyperventilation.

Answer (D) is correct (0488). *(AIM Para 602)*
   Hyperventilation occurs when an excessive amount of carbon dioxide is passed out of the body and too much oxygen is retained. This occurs when breathing rapidly.
   Answer (A) is incorrect because hypoxia results from a lack of oxygen (instead of too much) in the bloodstream. Answers (B) and (C) are incorrect because aerosinusitis and aerotitus are inflammations of the sinuses and the inner ear, respectively, caused by changes in atmospheric pressure.

2.
0490. Which would most likely result in hyperventilation?

A— Emotional tension, anxiety, or fear.
B— The excessive consumption of alcohol.
C— An extremely slow rate of breathing and insufficient oxygen.
D— An extreme case of relaxation or sense of well-being.

Answer (A) is correct (0490). *(AIM Para 602)*
   Hyperventilation usually occurs when one becomes excited or undergoes stress, which results in an increase in one's rate of breathing.
   Answer (B) is incorrect because hyperventilation is excessive oxygen, not alcohol, in the bloodstream. Answers (C) and (D) are incorrect because they are symptoms of hypoxia.

3.
0489. A pilot should be able to overcome the symptoms or avoid future occurrences of hyperventilation by

A— closely monitoring the flight instruments to control the airplane.
B— slowing the breathing rate, breathing into a bag, or talking aloud.
C— increasing the breathing rate in order to increase lung ventilation.
D— refraining from the use of over-the-counter remedies and drugs such as antihistamines, cold tablets, tranquilizers, etc.

Answer (B) is correct (0489). *(AIM Para 602)*
   To recover from hyperventilation, the pilot should slow the breathing rate, breathe into a bag, or simply talk aloud.
   Answer (A) is incorrect because closely monitoring the flight instruments is used to overcome vertigo (disorientation). Answer (C) is incorrect because increased breathing would aggravate hyperventilation. Answer (D) is incorrect because over-the-counter remedies and drugs, such as antihistamines, cold tablets, tranquilizers, etc., are not related to hyperventilation. They should be avoided by pilots.

Chapter 10: Flight Physiology and Flight Operations

SPATIAL DISORIENTATION

4.
0493. A pilot is more subject to spatial disorientation if

A— ignoring or overcoming the sensations of muscles and inner ear.
B— kinesthetic senses are ignored.
C— eyes are moved often in the process of cross-checking the flight instruments.
D— body signals are used to interpret flight attitude.

Answer (D) is correct (0493). *(FTH Chap 1)*
Spatial disorientation is a state of temporary confusion resulting from misleading information being sent to the brain by various sensory organs. Thus, the pilot should ignore sensations of muscles and inner ear and kinesthetic senses (those which sense motion).
Answers (A), (B), and (C) are incorrect because each helps overcome spatial disorientation.

5.
0494. A state of temporary confusion resulting from misleading information being sent to the brain by various sensory organs is defined as

A— spatial disorientation.
B— hyperventilation.
C— hypoxia.
D— motion sickness.

Answer (A) is correct (0494). *(FTH Chap 1)*
A state of temporary confusion resulting from misleading information being sent to the brain by various sensory organs is defined as vertigo (spatial disorientation). More simply, the pilot cannot determine his/her relationship to the earth's horizon.
Answer (B) is incorrect because hyperventilation causes excessive oxygen and/or a decrease in carbon dioxide in the bloodstream. Answer (C) is incorrect because hypoxia is a lack of oxygen to the brain. Answer (D) is incorrect because motion sickness, a form of nausea, may be one symptom of spatial disorientation.

CARBON MONOXIDE

6.
0491. Large accumulations of carbon monoxide in the human body result in

A— tightness across the forehead.
B— loss of muscular power.
C— an increased sense of well-being.
D— being too warm.

Answer (B) is correct (0491). *(AC 20-328)*
Carbon monoxide reduces the ability of the blood to carry oxygen. Large accumulations result in loss of muscular power.
Answer (A) is incorrect because it describes an early symptom, not the effect of large accumulations. Answer (C) is incorrect because euphoria is a result of the lack of sufficient oxygen. Answer (D) is incorrect because it describes a symptom of hyperventilation.

7.
0492. Susceptibility to carbon monoxide poisoning increases as

A— altitude increases.
B— altitude decreases.
C— air pressure increases.
D— humidity of the air decreases.

Answer (A) is correct (0492). *(AIM Para 601,603)*
Carbon monoxide poisoning results in an oxygen-deficiency. Since there is less oxygen available at higher altitudes, carbon monoxide poisoning can occur with lesser amounts of carbon monoxide.
Answers (B) and (C) are incorrect because there is more available oxygen at lower altitudes and at higher pressures. Answer (D) is incorrect because the humidity level is independent of the carbon monoxide and/or oxygen level.

PREFLIGHT INSPECTION

8.
0224. Prior to every flight, a pilot should at least

A— check the operation of the ELT.
B— drain fuel from each quick drain.
C— perform a walk-around inspection of the aircraft.
D— check the required documents aboard the aircraft.

Answer (C) is correct (0224). *(FTH Chap 5)*
 At a minimum, prior to every flight, pilots should perform a walk-around inspection of the airplane to ensure that there has been no damage on the ramp or other visible problems such as oil or fuel leaks.
 Answer (A) is incorrect because the ELT (emergency locator transmitter) is not routinely checked as part of the preflight inspection. It is inspected by a certified mechanic during the annual airplane inspection. Answer (B) is incorrect because draining fuel from each quick drain is appropriate for the first flight of the day, but not necessary on repeated flights. Answer (D) is incorrect because if the required documents were checked and no one else has used the airplane, there is no need to recheck them.

9.
0225. What special check should be made on an aircraft during preflight after it has been stored an extended period of time?

A— ELT batteries and operation.
B— Condensation in the fuel tanks.
C— Damage or obstructions caused by animals, birds, or insects.
D— Lubrication of control systems and proper inflation of landing gear struts.

Answer (C) is correct (0225). *(FTH Chap 5)*
 When airplanes have been stored for an extended period, it is possible that animals, birds, or insects have damaged or obstructed the airplane. Examples are bird nests within the engine cowling which would interfere with engine cooling or insect damage of pitot tubes.
 Answers (A), (B), and (D) are incorrect because checking ELT (emergency locator transmitter) batteries and operation, fuel tank condensation, lubrication of control systems, and landing gear struts are normal preflight checks, not special checks after long-term storage.

10.
0226. The use of a written checklist for preflight inspection and starting the engine is recommended

A— as an excellent crutch for those pilots with a faulty memory.
B— for memorizing the procedures in an orderly sequence.
C— as a procedure to instill confidence in the passengers.
D— to ensure that all necessary items are checked in a logical sequence.

Answer (D) is correct (0226). *(FTH Chap 5)*
 The purpose of a written checklist for preflight inspection and engine starting is to assure complete and orderly inspection of all necessary items. Without a checklist it is possible to omit some steps or perhaps do them out of a necessary order.
 Answers (A), (B), and (C) are incorrect because, while each is a by-product of written checklists, none is the primary purpose.

11.
0223. Is it necessary to preflight an aircraft that was hangared the night before in ready-to-fly condition?

A— Yes, because fuel contamination from condensation is possible.
B— Yes, because the oil level should always be checked.
C— No, if the same person who hangared the aircraft will act as pilot in command.
D— No, if the aircraft has not been handled since hangaring.

Answer (A) is correct (0223). *(FTH Chap 5)*
 Fuel contamination from condensation is possible and cannot be determined other than by draining fuel from the sumps of the tank to see if there is any water.
 Answer (B) is incorrect because, if the oil level was sufficient the night before and there are no visible leaks or drippings, there should be no problems. Answers (C) and (D) are incorrect because preflight is always necessary.

Chapter 10: Flight Physiology and Flight Operations

STARTING THE ENGINE

12.
0170. The pilot's first action after starting an aircraft engine should be to

A— adjust for proper RPM and check for desired indications on the engine gauges.
B— check the magneto or ignition switch momentarily in the OFF position for proper grounding.
C— test each brake and the parking brake.
D— visually clear the area for people and obstacles.

Answer (A) is correct (0170). (PHAK Chap II)
After the engine starts, the engine speed should be adjusted to the proper RPM. Then the engine gauges should be reviewed, with the oil pressure being the most important gauge initially.

Answer (B) is incorrect because turning off the magnetos will turn off the engine. Answer (C) is incorrect because the brakes should be on before starting the engine. Answer (D) is incorrect because one should visually clear the area for people and obstacles before starting the engine.

13.
0240. When starting an airplane engine by hand, it is extremely important that a competent pilot

A— call "contact" before touching the propeller.
B— be at the controls in the cockpit.
C— be in charge in the cockpit and call out all commands.
D— turn the propeller and call out all commands.

Answer (B) is correct (0240). (PHAK Chap II)
Because of the hazards involved in handstarting airplane engines, every precaution should be exercised. It is extremely important that a competent pilot be at the controls in the cockpit. Also, the person turning the propeller should be thoroughly familiar with the technique.

Answer (A) is incorrect because the person handpropping the airplane yells "gas on, switch off, throttle closed, brakes set" before touching the propeller. Answers (C) and (D) are incorrect because the person handpropping the airplane calls out the commands. Answer (D) is incorrect because a pilot should be at the controls of the airplane.

TAXIING TECHNIQUE

14.
0164. Which aileron position should a pilot generally use when taxiing during strong quartering headwinds?

A— Aileron up on the side from which the wind is blowing.
B— Aileron down on the side from which the wind is blowing.
C— Aileron parallel to the ground on the side from which the wind is blowing.
D— Neutral.

Answer (A) is correct (0164). (FTH Chap 5)
When there is a strong quartering headwind, the aileron should be up on the side from which the wind is blowing to help keep the wind from getting under that wing and blowing the aircraft over.

15.
0163. When taxiing during strong quartering tailwinds, which aileron position should be used?

A— Aileron up on the side from which the wind is blowing.
B— Aileron down on the side from which the wind is blowing.
C— Aileron parallel to the ground on the side from which the wind is blowing.
D— Neutral.

Answer (B) is correct (0163). (FTH Chap 5)
When there is a strong quartering tailwind, the aileron should be down on the side from which the wind is blowing (upwind not downwind side) to help keep the wind from getting under that wing and flipping the airplane over.

16.
0165. Which wind condition would be most critical when taxiing a nosewheel-equipped, high-wing airplane?

A— Direct headwind.
B— Direct crosswind.
C— Quartering headwind.
D— Quartering tailwind.

Answer (D) is correct (0165). *(FTH Chap 5)*
The most critical wind condition when taxiing a nosewheel-equipped high-wing airplane is a quartering tailwind, which can flip a high-wing airplane over on its top. This should be prevented by holding the elevator in the down position, i.e., controls forward, and the aileron down on the side from which the wind is coming.

17.
0159. (Refer to figure 14.) How should the flight controls be positioned while taxiing a tricycle-gear equipped airplane during a left quartering headwind as depicted by "A"?

A— Left aileron up, neutral elevator.
B— Left aileron down, neutral elevator.
C— Left aileron up, down elevator.
D— Left aileron down, down elevator.

Answer (A) is correct (0159). *(FTH Chap 5)*
Given a left quartering headwind, the left aileron should be kept up so the wind does not get under the left wing and flip the airplane over. The elevator should be neutral to keep from putting too much or too little weight on the nosewheel.

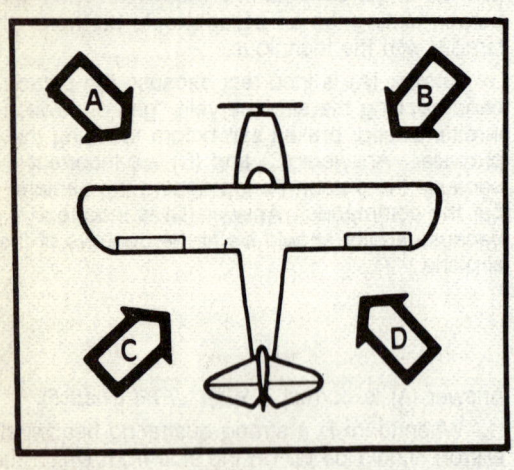

FIGURE 14.—Control Position for Taxi.

18.
0160. (Refer to figure 14.) How should the flight controls be positioned while taxiing a tricycle-gear equipped airplane during a left quartering tailwind as depicted by "C"?

A— Left aileron up, neutral elevator.
B— Left aileron down, neutral elevator.
C— Left aileron up, down elevator.
D— Left aileron down, down elevator.

Answer (D) is correct (0160). *(FTH Chap 5)*
With a left quartering tailwind, the left aileron should be down so the wind does not get under the left wing and flip the airplane over. Also, the elevator should be down, i.e., controls forward, so the wind does not get under the tail and blow the airplane tail over front.

19.
0161. (Refer to figure 14.) While taxiing a tricycle-gear equipped airplane during a right quartering headwind as depicted by "B," the right aileron should be held

A— up and the elevator neutral.
B— down and the elevator neutral.
C— up and the elevator down.
D— down and the elevator down.

Answer (A) is correct (0161). *(FTH Chap 5)*
When there is a right quartering headwind, the right aileron should be up so the wind does not get under the right wing and blow the airplane over. The elevator should be neutral to keep from putting too much or too little weight on the nosewheel.

Chapter 10: Flight Physiology and Flight Operations

20.
0162. (Refer to figure 14.) While taxiing a tricycle-gear equipped airplane during a right quartering tailwind as depicted by "D," the right aileron should be held

A— up and the elevator neutral.
B— down and the elevator neutral.
C— up and the elevator down.
D— down and the elevator down.

Answer (D) is correct (0162). *(FTH Chap 5)*

When there is a right quartering tailwind, the right aileron should be held down so the wind does not get under the right wing and flip the airplane over. Also, the elevator should be down, i.e., controls forward, so the wind does not get under the tail and blow the airplane tail over front.

AIRSPEED

21.
0230. If severe turbulence is encountered, the airplane's airspeed should be reduced to

A— maneuvering speed.
B— the minimum steady flight speed in the landing configuration.
C— normal operation speed.
D— maximum structural cruising speed.

Answer (A) is correct (0230). *(PHAK Chap V)*

In severe turbulence, the airplane should be reduced to its maneuvering speed, i.e., the airspeed at which the airplane will stall prior to receiving an abrupt load on the airplane sufficient to cause it structural damage.

Answer (B) is incorrect because the minimum steady flight speed in the landing configuration would be relatively close to stall speed. Thus, the airplane would be apt to stall in the changing wind gusts and directions associated with turbulence. Answers (C) and (D) are incorrect because normal operation speed and maximum structural cruising speed are greater than maneuvering speed and may result in structural damage to the airplane.

22.
0460. Upon encountering severe turbulence, which condition should the pilot attempt to maintain?

A— Constant altitude.
B— Constant airspeed (V_A).
C— Level flight attitude.
D— Constant altitude and constant airspeed.

Answer (C) is correct (0460). *(PHAK Chap I)*

In severe turbulence, the pilot should attempt to maintain a level flight attitude, which is simply "control of the airplane."

Answers (A) and (D) are incorrect because maintaining a constant altitude will require additional control movements, adding stress to the airplane. Answers (B) and (D) are incorrect because, in severe turbulence, the airspeed will fluctuate back and forth due to the wind shears and wind shifts that cause the turbulence.

23.
0248. The most important rule to remember in the event of a power failure after becoming airborne is to

A— quickly check the fuel supply for possible fuel exhaustion.
B— determine the wind direction to plan for the forced landing.
C— turn back immediately to the takeoff runway.
D— maintain safe airspeed.

Answer (D) is correct (0248). *(FTH Chap 11)*

In the event of a power failure after becoming airborne, the most important rule to remember is to maintain safe airspeed. Invariably, with a power failure, one returns to ground, but emphasis should be put on a controlled return rather than a crash return. Many pilots attempt to maintain altitude at the expense of airspeed, resulting in a stall or stall/spin.

Answers (A), (B), and (C) are incorrect because checking the fuel supply, determining wind direction for forced landing, and turning back to the takeoff runway are appropriate if one has sufficient altitude and time, but the first and foremost rule is to maintain safe airspeed.

24.
0243. After takeoff, which airspeed would permit the pilot to gain the most altitude in a given period of time?

A— Cruising climb speed.
B— Best rate-of-climb speed.
C— Best angle-of-climb speed.
D— Minimum control speed.

Answer (B) is correct (0243). *(FAR 1)*
The best rate-of-climb speed would give the most gain in altitude in a given time.

Answer (A) is incorrect because cruising climb speed produces a "leisurely" climb rate. Answer (C) is incorrect because the best angle of climb would give the most gain in altitude in the shortest distance. Answer (D) is incorrect because minimum control airspeed is just above stall speed and wastes altitude.

25.
0244. Which would provide the greatest gain in altitude in the shortest distance during climb after takeoff?

A— Steepest pitch attitude.
B— Cruising climb speed.
C— Best rate-of-climb speed.
D— Best angle-of-climb speed.

Answer (D) is correct (0244). *(FAR 1)*
The best angle-of-climb speed would give the most gain in altitude for the shortest distance.

Answer (A) is incorrect because the steepest pitch altitude will produce an airspeed below V_x (best angle-of-climb speed) which will decrease until a stall occurs. Answer (B) is incorrect because cruising climb speed produces a "leisurely" climb rate. Answer (C) is incorrect because the best rate of climb speed provides the greatest gain in altitude in the shortest time.

COLLISION AVOIDANCE

26.
0284. How can you determine if another aircraft is on a collision course with your aircraft?

A— The other aircraft will be pointed directly at your aircraft.
B— The other aircraft will always appear to get larger and closer at a rapid rate.
C— The nose of each aircraft is pointed at the same point in space.
D— There will be no apparent relative motion between your aircraft and the other aircraft.

Answer (D) is correct (0284). *(AIM Para 607)*
Any aircraft that appears to have no relative motion and stays in one scan quadrant is likely to be on a collision course. Also, if a target shows no lateral or vertical motion, but increases in size, take evasive action.

Answer (A) is incorrect because an aircraft may be momentarily pointed at your aircraft during a climb or turn and pose no collision threat. Answer (B) is incorrect because aircraft on collision courses may not always appear to grow larger or to close at a rapid rate. Frequently, the degree of proximity cannot be detected. Answer (C) is incorrect because you may not be able to tell in exactly which direction the other airplane is pointed. Even if you could determine the direction of the other airplane, you may not be able to accurately project the flightpaths of the two airplanes to determine if they indeed point to the same point in space and will arrive there at the same time (i.e., collide).

27.
0285. Prior to starting each maneuver, a pilot should

A— check altitude, airspeed, and heading indications.
B— visually scan the area below your aircraft in case of an accidental stall or spin.
C— advise ATC of your intentions so as to receive timely traffic advisories.
D— visually scan the entire area for collision avoidance.

Answer (D) is correct (0285). *(AC 90-48C)*
Prior to each maneuver, a pilot should visually scan the entire area for collision avoidance. Many maneuvers require a clearing turn which fills this purpose.

Answer (A) is incorrect because altitude, speed, and heading may not all be critical to every maneuver. Collision avoidance is! Answer (B) is incorrect because all directions (above, behind, and to each side) as well as below should be scanned. Answer (C) is incorrect because VFR pilots are not always in contact with ATC.

FAA LISTING OF
SUBJECT MATTER KNOWLEDGE CODES

The next six pages reprint the FAA's subject matter codes. These are the codes that will appear on your *Airman Written Test Report* (FAA Form 8080-2). See the illustration on page 12. The FAA will list the subject matter code of each question answered incorrectly on your written test.

When you receive your Form 8080-2 from the FAA, you can trace the subject matter codes listed on it to the next six pages to find out which topics you had difficulty with. You should discuss your written test results with your CFI.

DEPARTMENT OF TRANSPORTATION

FEDERAL AVIATION ADMINISTRATION

SUBJECT MATTER KNOWLEDGE CODES

To determine the knowledge area in which a particular question was incorrectly answered, compare the subject matter code(s) on AC Form 8080-2, Airmen Written Test Report, to the subject matter outline that follows. The total number of test items missed may differ from the number of subject matter codes shown on the AC Form 8080-2, since you may have missed more than one question in a certain subject matter code.

FAR 1   **Definitions and Abbreviations**

- A01   General Definitions
- A02   Abbreviations and Symbols

FAR 23   **Airworthiness Standards: Normal, Utility, and Acrobatic Category Aircraft**

- A10   General

FAR 43   **Maintenance, Preventive Maintenance, Rebuilding and Alteration**

- A15   General
- A16   Appendices

FAR 61   **Certification: Pilots and Flight Instructors**

- A20   General
- A21   Aircraft Ratings and Special Certificates
- A22   Student Pilots
- A23   Private Pilots
- A24   Commercial Pilots
- A25   Airline Transport Pilots
- A26   Flight Instructors
- A27   Appendix A: Practical Test Requirements for Airline Transport Pilot Certificates and Associated Class and Type Ratings
- A28   Appendix B: Practical Test Requirements for Rotorcraft Airline Transport Pilot Certificates with a Helicopter Class Rating and Associated Type Ratings
- A29   Recreational Pilot

FAR 63   **Certification: Flight Crewmembers Other Than Pilots**

- A30   General
- A31   Flight Engineers
- A32   Flight Navigators

FAR 65   **Certification: Airmen Other Than Flight Crewmembers**

- A40   General
- A41   Aircraft Dispatchers

FAR 67   **Medical Standards and Certification**

- A50   General
- A51   Certification Procedures

FAR 71   **Designation of Federal Airways, Area Low Routes, Controlled Airspace, and Reporting Points**

- A60   General
- A61   Airport Radar Service Areas

FAR 73   **Special Use Airspace**

- A70   General
- A71   Restricted Areas
- A72   Prohibited Areas

FAR 75   **Establishment of Jet Routes and Area High Routes**

- A80   General

FAR 77   **Objects Affecting Navigable Airspace**

- A90   General

FAR 91   **General Operating Rules**

- B01   General
- B02   Flight Rules
- B03   Maintenance, Preventive Maintenance, and Alterations
- B04   Large and Turbine-Powered Multiengine Airplanes
- B05   Operating Noise Limits
- B06   Appendix A: Category II Operations Manual, Instruments, Equipment, and Maintenance

FAR 103   **Ultralight Vehicles**

- C01   General
- C02   Operating Rules

FAR 108   **Airplane Operator Security**

- C10   General

FAR 121   **Certification and Operations: Domestic, Flag and Supplemental Air Carriers and Commercial Operators of Large Aircraft**

- D01   General
- D02   Certification Rules for Domestic and Flag Air Carriers
- D03   Certification Rules for Supplemental Air Carriers and Commercial Operators
- D04   Rules Governing all Certificate Holders Under This Part

FAA Subject Matter Knowledge Codes

D05	Approval of Routes: Domestic and Flag Air Carriers	E12	Special Federal Aviation Regulations SFAR No. 36
D06	Approval of Areas and Routes for Supplemental Air Carriers and Commercial Operators	E13	Special Federal Aviation Regulations SFAR No. 38
D07	Manual Requirements	**FAR 143**	**Certification: Ground Instructors**
D08	Aircraft Requirements	F01	General
D09	Airplane Performance Operating Limitations	**US HMR 172**	**Hazardous Materials Table**
D10	Special Airworthiness Requirements	F02	General
D11	Instrument and Equipment Requirements		
D12	Maintenance, Preventive Maintenance, and Alterations	**US HMR 175**	**Materials Transportation Bureau Hazardous Materials Regulations (HMR)**
D13	Airman and Crewmember Requirements		
D14	Training Program	G01	General Information and Regulations
D15	Crewmember Qualifications	G02	Loading, Unloading, and Handling
D16	Aircraft Dispatcher Qualifications and Duty Time Limitations: Domestic and Flag Air Carriers	G03	Specific Regulation Applicable According to Classification of Material
D17	Flight Time Limitations and Rest Requirements: Domestic Air Carriers	**NTSB 830**	**Rules Pertaining to the Notification and Reporting of Aircraft Accidents or Incidents and Overdue Aircraft, and Preservation of Aircraft Wreckage, Mail, Cargo, and Records**
D18	Flight Time Limitations: Flag Air Carriers		
D19	Flight Time Limitations: Supplemental Air Carriers and Commercial Operators		
D20	Flight Operations		
D21	Dispatching and Flight Release Rules		
D22	Records and Reports	G10	General
D23	Crewmember Certificate: International	G11	Initial Notification of Aircraft Accidents, Incidents, and Overdue Aircraft
D24	Special Federal Aviation Regulation SFAR No. 14	G12	Preservation of Aircraft Wreckage, Mail, Cargo, and Records
FAR 125	**Certification and Operations: Airplanes Having a Seating Capacity of 20 or More Passengers or a Maximum Payload Capacity of 6,000 Pounds or More**	G13	Reporting of Aircraft Accidents, Incidents, and Overdue Aircraft
		AC 61-23	**Pilot's Handbook of Aeronautical Knowledge**
D30	General	H01	Principles of Flight
D31	Certification Rules and Miscellaneous Requirements	H02	Airplanes and Engines
D32	Manual Requirements	H03	Flight Instruments
D33	Airplane Requirements	H04	Airplane Performance
D34	Special Airworthiness Requirements	H05	Weather
D35	Instrument and Equipment Requirements	H06	Basic Calculations Using Navigational Computers or Electronic Calculators
D36	Maintenance	H07	Navigation
D37	Airman and Crewmember Requirements	H08	Flight Information Publications
D38	Flight Crewmember Requirements	H09	Appendix 1: Obtaining FAA Publications
D39	Flight Operations		
D40	Flight Release Rules	**AC 91-23**	**Pilot's Weight and Balance Handbook**
D41	Records and Reports	H10	Weight and Balance Control
FAR 135	**Air Taxi Operators and Commercial Operators**	H11	Terms and Definitions
		H12	Empty Weight Center of Gravity
E01	General	H13	Index and Graphic Limits
E02	Flight Operations	H14	Change of Weight
E03	Aircraft and Equipment	H15	Control of Loading — General Aviation
E04	VFR/IFR Operating Limitations and Weather Requirements	H16	Control of Loading — Large Aircraft
E05	Flight Crewmember Requirements	**AC 60-14**	**Aviation Instructor's Handbook**
E06	Flight Crewmember Flight Time Limitations and Rest Requirements	H20	The Learning Process
E07	Crewmember Testing Requirements	H21	Human Behavior
E08	Training	H22	Effective Communication
E09	Airplane Performance Operating Limitations	H23	The Teaching Process
E10	Maintenance, Preventive Maintenance, and Alterations	H24	Teaching Methods
E11	Appendix A: Additional Airworthiness Standards for 10 or More Passenger Airplanes	H25	The Instructor as a Critic
		H26	Evaluation

197

H27	Instructional Aids	I04	Basic Flight Instruments
H30	Flight Instructor Characteristics and Responsibilities	I05	Attitude Instrument Flying — Airplanes
		I06	Attitude Instrument Flying — Helicopters
H31	Techniques of Flight Instruction	I07	Electronic Aids to Instrument Flying
H32	Planning Instructional Activity	I08	Using the Navigation Instruments
H40	Aircraft Maintenance Instructor Characteristics and Responsibilities	I09	Radio Communications Facilities and Equipment
H41	Integrated Job Training	I10	The Federal Airways System and Controlled Airspace
H42	Planning Aircraft Maintenance Instructional Activities	I11	Air Traffic Control
		I12	ATC Operations and Procedures
H43	Appendix	I13	Flight Planning

AC 61-21 Flight Training Handbook

I14	Appendix: Instrument Instructor Lesson Guide — Airplanes
I15	Segment of En Route Low Altitude Chart

H50	Introduction to Flight Training
H51	Introduction to Airplanes and Engines
H52	Introduction to the Basics of Flight
H53	The Effect and Use of Controls
H54	Ground Operations
H55	Basic Flight Maneuvers
H56	Airport Traffic Patterns and Operations
H57	Takeoffs and Departure Climbs
H58	Landing Approaches and Landings
H59	Faulty Approaches and Landings
H60	Proficiency Flight Maneuvers
H61	Cross-Country Flying
H62	Emergency Flight by Reference to Instruments
H63	Night Flying
H64	Seaplane Operations
H65	Transition to Other Airplanes
H66	Principles of Flight and Performance Characteristics

AC 00-6 Aviation Weather

I20	The Earth's Atmosphere
I21	Temperature
I22	Atmospheric Pressure and Altimetry
I23	Wind
I24	Moisture, Cloud Formation, and Precipitation
I25	Stable and Unstable Air
I26	Clouds
I27	Air Masses and Fronts
I28	Turbulence
I29	Icing
I30	Thunderstorms
I31	Common IFR Producers
I32	High Altitude Weather
I33	Arctic Weather
I34	Tropical Weather
I35	Soaring Weather
I36	Glossary of Weather Terms

AC 61-13 Basic Helicopter Handbook

AC 00-45 Aviation Weather Services

H70	General Aerodynamics
H71	Aerodynamics of Flight
H72	Loads and Load Factors
H73	Function of the Controls
H74	Other Helicopter Components and Their Functions
H75	Introduction to the Helicopter Flight Manual
H76	Weight and Balance
H77	Helicopter Performance
H78	Some Hazards of Helicopter Flight
H79	Precautionary Measures and Critical Conditions
H80	Helicopter Flight Maneuvers
H81	Confined Area, Pinnacle, and Ridgeline Operations
H82	Glossary

I40	The Aviation Weather Service Program
I41	Surface Aviation Weather Reports
I42	Pilot and Radar Reports and Satellite Pictures
I43	Aviation Weather Forecasts
I44	Surface Analysis Chart
I45	Weather Depiction Chart
I46	Radar Summary Chart
I47	Significant Weather Prognostics
I48	Winds and Temperatures Aloft
I49	Composite Moisture Stability Chart
I50	Severe Weather Outlook Chart
I51	Constant Pressure Charts
I52	Tropopause Data Chart
I53	Tables and Conversion Graphs

Gyroplane Flight Training Manual

AIM Airman's Information Manual

H90	Gyroplane Systems
H91	Gyroplane Terms
H92	Use of Flight Controls (Gyroplane)
H93	Fundamental Maneuvers of Flight (Gyroplane)
H94	Basic Flight Maneuvers (Gyroplane)

J01	Air Navigation Radio Aids
J02	Radar Services and Procedures
J03	Airport Lighting Aids
J04	Air Navigation and Obstruction Lighting
J05	Airport Marking Aids
J06	Airspace — General
J07	Uncontrolled Airspace
J08	Controlled Airspace
J09	Special Use Airspace
J10	Other Airspace Areas
J11	Service Available to Pilots

AC 61-27 Instrument Flying Handbook

I01	Training Considerations
I02	Instrument Flying: Coping with Illusions in Flight
I03	Aerodynamic Factors Related to Instrument Flying

Code	Description
J12	Radio Communications Phraseology and Techniques
J13	Airport Operations
J14	ATC Clearance/Separations
J15	Preflight
J16	Departure Procedures
J17	En Route Procedures
J18	Arrival Procedures
J19	Pilot/Controller Roles and Responsibilities
J20	National Security and Interception Procedures
J21	Emergency Procedures — General
J22	Emergency Services Available to Pilots
J23	Distress and Urgency Procedures
J24	Two-Way Radio Communications Failure
J25	Meteorology
J26	Altimeter Setting Procedures
J27	Wake Turbulence
J28	Bird Hazards, and Flight Over National Refuges, Parks, and Forests
J29	Potential Flight Hazards
J30	Safety, Accident, and Hazard Reports
J31	Fitness for Flight
J32	Type of Charts Available
J33	Pilot Controller Glossary
J34	Airport/Facility Directory
J35	En Route Low Altitude Chart
J36	En Route High Altitude Chart
J37	Sectional Chart
J38	WAC Chart
J39	Terminal Area Chart
J40	Standard Instrument Departure (SID) Chart
J41	Standard Terminal Arrival (STAR) Chart
J42	Instrument Approach Procedures
J43	Helicopter Route Chart

AC 67-2   Medical Handbook for Pilots

Code	Description
J50	The Flyer's Environment
J51	The Pressure is On
J52	Hypoxia
J53	Hyperventilation
J54	The Gas in the Body
J55	The Ears
J56	Alcohol
J57	Drugs and Flying
J58	Carbon Monoxide
J59	Vision
J60	Night Flight
J61	Cockpit Lighting
J62	Disorientation (Vertigo)
J63	Motion Sickness
J64	Fatigue
J65	Noise
J66	Age
J67	Some Psychological Aspects of Flying
J68	The Flying Passenger

ADDITIONAL ADVISORY CIRCULARS

Code	Description
K01	AC 00-24, Thunderstorms
K02	AC 00-30, Rules of Thumb for Avoiding or Minimizing Encounters with Clear Air Turbulence
K03	AC 00-34, Aircraft Ground Handling and Servicing
K04	AC 00-50, Low Level Wind Shear
K10	AC 20-5, Plane Sense
K11	AC 20-29, Use of Aircraft Fuel Anti-Icing Additives
K12	AC 20-32, Carbon Monoxide (CO) Contamination in Aircraft — Detection and Prevention
K13	AC 20-43, Aircraft Fuel Control
K14	AC 20-64, Maintenance Inspection Notes for Lockheed L-188 Series Aircraft
K15	AC 20-76, Maintenance Inspection Notes for Boeing B-707/720 Series Aircraft
K16	AC 20-78, Maintenance Inspection Notes for McDonnell Douglas DC-8 Series Aircraft
K17	AC 20-84, Maintenance Inspection Notes for Boeing B-727 Series Aircraft
K18	AC 20-99, Antiskid and Associated Systems
K19	AC 20-101, Airworthiness Approval of Omega/VLF Navigation Systems for the United States NAS and Alaska
K20	AC 20-103, Aircraft Engine Crankshaft Failure
K21	AC 20-106, Aircraft Inspection for the General Aviation Aircraft Owner
K22	AC 20-113, Pilot Precautions and Procedures to be Taken in Preventing Aircraft Reciprocating Engine Induction System and Fuel System Icing Problems
K23	AC 20-121, Airworthiness Approval of Airborne Loran-C Navigation Systems for Use in the U.S. National Airspace System
K24	AC 20-125, Water in Aviation Fuels
K30	AC 23.679-1, Control System Locks
K31	AC 23.1521-1, Approval of Automobile Gasoline (Autogas) in Lieu of Aviation Gasoline (Avgas) in Small Airplanes with Reciprocating Engines
K40	AC 25-4, Inertial Navigation Systems (INS)
K41	AC 25.253-1, High-Speed Characteristics
K50	AC 29-2, Certification of Transportation Category Rotorcraft
K60	AC 33.65-1, Surge and Stall Characteristics of Aircraft Turbine Engines
K70	AC 43-9, Maintenance Records
K71	AC 43-12, Preventive Maintenance
K80	AC 60-4, Pilot's Spatial Disorientation
K81	AC 60-6, Airplane Flight Manuals (AFM), Approved Manual Materials, Markings, and Placards — Airplanes
K90	AC 60-12, Availability of Industry-Developed Guidelines for the Conduct of the Biennial Flight Review
L01	AC 61-9, Pilot Transition Courses for Complex Single-Engine and Light, Twin-Engine Airplanes
L02	AC 61-10, Private and Commercial Pilots Refresher Courses
L03	AC 61-12, Student Pilot Guide
L04	AC 61-47, Use of Approach Slope Indicators for Pilot Training
L05	AC 61-65, Certification: Pilot and Flight Instructors
L06	AC 61-66, Annual Pilot In Command Proficiency Checks

Code	Description
L07	AC 61-67, Hazards Associated with Spins in Airplanes Prohibited from Intentional Spinning
L08	AC 61-84, Role of Preflight Preparation
L09	AC 61-89, Pilot Certificates: Aircraft Type Ratings
L10	AC 61-92, Use of Distractions During Pilot Certification Flight Tests
L11	AC 61-94, Pilot Transition Course for Self-Launching or Powered Sailplanes (Motorgliders)
L20	AC 67-1, Medical Information for Air Ambulance Operators
L30	AC 90-23, Aircraft Wake Turbulence
L31	AC 90-34, Accidents Resulting from Wheelbarrowing in Tricycle-Gear Equipped Aircraft
L32	AC 90-42, Traffic Advisory Practices at Nontower Airports
L33	AC 90-45, Approval of Area Navigation Systems for Use in the U.S. National Airspace System
L34	AC 90-48, Pilots' Role in Collision Avoidance
L35	AC 90-58, VOR Course Errors Resulting from 50 kHz Channel Mis-Selection
L36	AC 90-66, Recommended Standard Traffic Patterns for Airplane Operations at Uncontrolled Airports
L37	AC 90-67, Light Signals from the Control Tower for Ground Vehicles, Equipment, and Personnel
L38	AC 90-79, Recommended Practices and Procedures for the Use of Electronic Long-Range Navigation Equipment
L39	AC 90-82, Random Area Navigation Routes
L40	AC 90-83, Terminal Control Areas (TCA)
L41	AC 90-85, Severe Weather Avoidance Plan (SWAP)
L42	AC 90-87, Helicopter Dynamic Rollover
L43	AC 90-88, Airport Radar Service Area (ARSA)
L50	AC 91-6, Water, Slush, and Snow on the Runway
L51	AC 91-8, Use of Oxygen by Aviation Pilots/Passengers
L52	AC 91-13, Cold Weather Operation of Aircraft
L53	AC 91-14, Altimeter Setting Sources
L54	AC 91-16, Category II Operations — General Aviation Airplanes
L55	AC 91-32, Safety In and Around Helicopters
L56	AC 91-42, Hazards of Rotating Propeller and Helicopter Rotor Blades
L57	AC 91-43, Unreliable Airspeed Indications
L58	AC 91-44, Operational and Maintenance Practices for Emergency Locator Transmitters and Receivers
L59	AC 91-46, Gyroscopic Instruments — Good Operating Practices
L60	AC 91-49, General Aviation Procedures for Flight in North Atlantic Minimum Navigation Performance Specifications Airspace
L61	AC 91-50, Importance of Transponder Operation and Altitude Reporting
L62	AC 91-51, Airplane Deice and Anti-Ice Systems
L63	AC 91-53, Noise Abatement Departure Profile
L64	AC 91-55, Reduction of Electrical System Failures Following Aircraft Engine Starting
L65	AC 91-58, Use of Pyrotechnic Visual Distress Signaling Devices in Aviation
L67	AC 91.83-1, Canceling or Closing Flight Plans
L70	AC 97-1, Runway Visual Range (RVR)
L80	AC 103-4, Hazard Associated with Sublimation of Solid Carbon Dioxide (Dry Ice) Aboard Aircraft
L81	AC 103-6, Ultralight Vehicle Operations Airports, ATC, and Weather
L82	AC 103-7, The Ultralight Vehicle
M01	AC 120-12, Private Carriage Versus Common Carriage of Persons or Property
M02	AC 120-27, Aircraft Weight and Balance Control
M03	AC 120-29, Criteria for Approving Category I and Category II Landing Minima for FAR 121 Operators
M04	AC 120-32, Air Transportation of Handicapped Persons
M10	AC 121-6, Portable Battery-Powered Megaphones
M11	AC 121-24, Passenger Safety Information Briefing and Briefing Cards
M12	AC 121-25, Additional Weather Information: Domestic and Flag Air Carriers
M13	AC 121-195, Alternate Operational Landing Distances for Wet Runways; Turbojet Powered Transport Category Airplanes
M20	AC 125-1, Operations of Large Airplanes Subject to Federal Aviation Regulations Part 125
M30	AC 135-3, Air Taxi Operators and Commercial Operators
M31	AC 135-9, FAR Part 135 Icing Limitations
M32	AC 135-12, Passenger Information, FAR Part 135: Passenger Safety Information Briefing and Briefing Cards
M40	AC 150/5345-28, Precision Approach Path Indicator (PAPI) Systems
M50	AC 20-34, Prevention of Retractable Landing Gear Failures
M51	AC 20-117, Hazards Following Ground Deicing and Ground Operations in Conditions Conducive to Aircraft Icing

American Soaring Handbook — Gliders

Code	Description
N01	A History of American Soaring
N02	Training
N03	Ground Launch
N04	Airplane Tow
N05	Meteorology
N06	Cross-Country and Wave Soaring
N07	Instruments and Oxygen
N08	Radio, Rope, and Wire
N09	Aerodynamics
N10	Maintenance and Repair

Soaring Flight Manual — Gliders

- N20 Sailplane Aerodynamics
- N21 Performance Considerations
- N22 Flight Instruments
- N23 Weather for Soaring
- N24 Medical Factors
- N25 Flight Publications and Airspace
- N26 Aeronautical Charts and Navigation
- N27 Computations for Soaring
- N28 Personal Equipment
- N29 Preflight and Ground Operations
- N30 Aerotow Launch Procedures
- N31 Ground Launch Procedures
- N32 Basic Flight Maneuvers and Traffic Patterns
- N33 Soaring Techniques
- N34 Cross-Country Soaring

Taming The Gentle Giant — Balloons

- O01 Design and Construction of Balloons
- O02 Fuel Source and Supply
- O03 Weight and Temperature
- O04 Flight Instruments
- O05 Balloon Flight Tips
- O06 Glossary

Balloon Federation Of America — Flight Instructor Manual

- O10 Flight Instruction Aids
- O11 Human Behavior and Pilot Proficiency
- O12 The Flight Check and the Designated Examiner

The Balloon Federation of America Handbook — Propane Systems

- O20 Propane Glossary
- O21 Chemical and Physical Systems
- O22 Cylinders
- O23 Lines and Fittings
- O24 Valves
- O25 Regulators
- O26 Burners
- O27 Propane Systems — Schematics
- O28 Propane References

The Balloon Federation of America Handbook — Avoiding Powerline Accidents

- O30 Excerpts

Balloon Flight Manual

- O40 Excerpts

Airship Operations Manual

- P01 Buoyancy
- P02 Aerodynamics
- P03 Free Ballooning
- P04 Aerostatics
- P05 Envelope
- P06 Car
- P07 Powerplant
- P08 Airship Ground Handling
- P09 Operating Instructions
- P10 History

International Flight Information Manual

- Q01 Passport and Visa
- Q02 International NOTAM Availability and Distribution
- Q03 National Security
- Q04 International Interception Procedures
- Q05 Intercept Pattern for Identification of Transport Aircraft
- Q06 Flight Planning Notes
- Q07 North Atlantic Minimum Navigation Requirements
- Q08 North American Routes for North Atlantic Traffic
- Q09 U.S. Aeronautical Telecommunications Services
- Q10 Charts and Publications for Flights Outside the U.S.
- Q11 Oceanic Long-Range Navigation Information

Aerodynamics For Naval Aviators, NAVWEPS 00-80T-80

- R01 Wing and Airfoil Forces
- R02 Planform Effects and Airplane Drag
- R10 Required Thrust and Power
- R11 Available Thrust and Power
- R12 Items of Airplane Performance
- R21 General Concepts and Supersonic Flow Patterns
- R22 Configuration Effects
- R31 Definitions
- R32 Longitudinal Stability and Control
- R33 Directional Stability and Control
- R34 Lateral Stability and Control
- R35 Miscellaneous Stability Problems
- R40 General Definitions and Structural Requirements
- R41 Aircraft Loads and Operating Limitations
- R50 Application of Aerodynamics to Specific Problems of Flying

Transport Airplane Operations Manual

- S01 Reserved
- T01 Reserved
- U01 Reserved
- V01 Reserved
- W01 Reserved
- X01 Reserved
- Y01 Reserved
- Z01 Visual Flight Rules Chart Users Guide
- Z02 Pilot's Guide to IVRS

NOTE: Most of the references and study materials listed in these subject matter knowledge codes are available through government outlets such as U.S. Government Printing Office bookstores. AC 00-2, Advisory Circular Checklist, transmits the status of all FAA advisory circulars (AC's), as well as FAA internal publications and miscellaneous flight information such as AIM, Airport/Facility Directory, written test question books, practical test standards, and other material directly related to a certificate or rating. To obtain a free copy of the AC 00-2, send your request to:

U.S. Department of Transportation
Utilization and Storage Section, M-443.2
Washington, DC 20590

CROSS-REFERENCES TO THE
FAA WRITTEN TEST QUESTION NUMBERS

The next three pages contain a listing of the FAA Recreational Pilot question numbers appearing in FAA-T-8080-14 (the FAA book of questions from which you will be taking your written test). The questions are numbered 0001 to 0500 in the FAA book. To the right of each FAA question number, we have added our chapter and question number. For example, the FAA's question 0001 is cross-referenced to our book as 4-25, which means it appears in Chapter 4 as question 25.

If no chapter/question number appears next to the FAA question number, the question is not applicable to the airplane rating written test, e.g., helicopter, glider, airship, etc. One question is a duplicate and is cross-referenced as such.

The first line of each of our answer explanations in Chapters 2 through 10 contains:

1. The correct answer,
2. The FAA question number, and
3. A reference for the answer explanation, e.g., an FAR number.

Thus, our questions are cross-referenced throughout this book to the FAA question numbers, and the next three pages cross-reference the FAA question numbers back to this book.

FAA Q. No.	Our Chap/ Q. No.	FAA Q. No.	Our Chap/ Q. No.	FAA Q. No.	Our Chap/ Q. No.	FAA Q. No.	Our Chap/ Q. No.
0001	4-25	0017	3-4	0033	3-28	0049	3-18
0002	4-24	0018	4-39	0034	3-29	0050	3-30
0003	4-26	0019	4-34	0035	4-33	0051	3-31
0004	4-27	0020	3-7	0036		0052	3-32
0005	4-32	0021	3-6	0037		0053	3-33
0006	3-5	0022	3-8	0038		0054	3-34
0007	3-3	0023	3-9	0039		0055	3-35
0008	3-2	0024	3-10	0040		0056	3-36
0009	4-31	0025	3-11	0041	3-19	0057	3-37
0010	4-29	0026	3-12	0042	3-20	0058	3-38
0011	4-30	0027	3-13	0043	3-21	0059	3-39
0012	4-38	0028	3-14	0044	3-22	0060	
0013	4-37	0029	3-24	0045	3-23	0061	
0014	4-28	0030	3-25	0046	3-15	0062	
0015	4-35	0031	3-26	0047	3-16	0063	
0016	4-36	0032	3-27	0048	3-17	0064	

(continued on next page)

Cross-References to the FAA Written Test Question Numbers

FAA Q. No.	Our Chap/ Q. No.	FAA Q. No.	Our Chap/ Q. No.	FAA Q. No.	Our Chap/ Q. No.	FAA Q. No.	Our Chap/ Q. No.
0065	4-51	0104	4-14	0143	2-17	0182	
0066	4-52	0105	4-16	0144	2-7	0183	
0067	4-54	0106	4-17	0145		0184	
0068	4-58	0107	4-19	0146		0185	
0069	4-59	0108	4-18	0147		0186	
0070	4-53	0109	4-21	0148		0187	6-10
0071	4-57	0110	4-20	0149		0188	6-11
0072	4-50	0111	4-8	0150		0189	6-12
0073	4-49	0112	2-4	0151		0190	6-15
0074	4-42	0113	2-5	0152		0191	6-13
0075	4-40	0114	2-19	0153		0192	6-14
0076	4-41	0115	2-1	0154		0193	6-16
0077	4-43	0116	2-8	0155		0194	
0078	4-68	0117	2-6	0156		0195	
0079	4-63	0118	2-28	0157	6-2	0196	
0080	4-64	0119	2-29	0158	6-1	0197	9-11
0081	4-65	0120	2-30	0159	10-17	0198	9-12
0082	4-66	0121	2-31	0160	10-18	0199	9-13
0083	4-69	0122	2-27	0161	10-19	0200	9-14
0084	4-70	0123	2-26	0162	10-20	0201	9-15
0085	4-67	0124	2-3	0163	10-15	0202	9-16
0086	4-61	0125	2-2	0164	10-14	0203	9-17
0087	4-62	0126	2-33	0165	10-16	0204	9-18
0088	4-60	0127	2-34	0166		0205	9-19
0089		0128	2-32	0167		0206	9-20
0090		0129	2-20	0168		0207	9-21
0091	4-1	0130	2-21	0169		0208	9-22
0092	4-3	0131	2-22	0170	10-12	0209	9-23
0093	4-2	0132	2-23	0171	6-3	0210	9-8
0094	4-5	0133	2-24	0172	6-4	0211	9-24
0095	4-4	0134	2-14	0173		0212	9-25
0096	4-6	0135	2-25	0174	6-5	0213	9-26
0097	4-11	0136	(Dupl. of 0123)	0175	6-6	0214	9-27
0098	4-9	0137	2-9	0176	6-7	0215	9-28
0099	4-7	0138	2-10	0177	6-8	0216	9-29
0100	4-10	0139	2-11	0178	6-9	0217	9-30
0101	4-12	0140	2-15	0179		0218	9-31
0102	4-13	0141	2-16	0180		0219	9-32
0103	4-15	0142	2-18	0181		0220	9-33

FAA Q. No.	Our Chap/ Q. No.	FAA Q. No.	Our Chap/ Q. No.	FAA Q. No.	Our Chap/ Q. No.	FAA Q. No.	Our Chap/ Q. No.
0221	9-34	0260	5-19	0299	8-17	0338	8-3
0222		0261	5-20	0300	8-29	0339	8-45
0223	10-11	0262	5-6	0301	8-31	0340	8-46
0224	10-8	0263	5-3	0302	8-7	0341	8-47
0225	10-9	0264	5-4	0303	8-8	0342	8-48
0226	10-10	0265	5-5	0304	8-9	0343	8-49
0227	4-46	0266	5-13	0305	8-92	0344	8-50
0228	4-47	0267	5-11	0306	8-96	0345	8-51
0229	4-48	0268	5-12	0307	8-97	0346	8-55
0230	10-21	0269	5-1	0308	8-95	0347	8-57
0231		0270	5-2	0309	8-94	0348	8-56
0232		0271	5-29	0310	8-93	0349	8-53
0233		0272	8-69	0311	8-98	0350	8-52
0234		0273	9-7	0312	9-1	0351	8-54
0235		0274	9-9	0313	5-26	0352	8-59
0236		0275	9-10	0314	5-31	0353	8-58
0237		0276	5-27	0315	5-32	0354	8-62
0238		0277	5-28	0316	8-4	0355	8-63
0239		0278	5-30	0317	8-6	0356	8-60
0240	10-13	0279	5-16	0318	8-11	0357	8-61
0241	3-41	0280	5-17	0319	8-5	0358	8-66
0242	3-40	0281	9-5	0320	8-12	0359	8-67
0243	10-24	0282	9-6	0321	8-10	0360	8-64
0244	10-25	0283	5-34	0322	8-13	0361	8-65
0245	3-42	0284	10-26	0323	8-32	0362	4-23
0246	4-44	0285	10-27	0324	8-33	0363	4-22
0247	4-45	0286	8-15	0325	8-34	0364	5-24
0248	10-23	0287	8-20	0326	8-16	0365	5-25
0249		0288	8-14	0327	8-35	0366	
0250	5-7	0289	8-18	0328	8-37	0367	5-14
0251	5-8	0290	8-19	0329	8-39	0368	5-33
0252	5-9	0291	8-21	0330	8-38	0369	8-68
0253	5-10	0292	8-23	0331	8-36	0370	8-79
0254	5-15	0293	8-22	0332	8-40	0371	8-75
0255		0294	8-30	0333	8-41	0372	8-78
0256	5-21	0295	8-24	0334	8-42	0373	8-80
0257	5-22	0296	8-25	0335	8-44	0374	8-72
0258	5-23	0297	8-28	0336	8-43	0375	8-81
0259	5-18	0298	8-26	0337		0376	8-77

Cross-References to the FAA Written Test Question Numbers

FAA Q. No.	Our Chap/Q. No.	FAA Q. No.	Our Chap/Q. No.	FAA Q. No.	Our Chap/Q. No.	FAA Q. No.	Our Chap/Q. No.
0377	8-27	0408	7-25	0439	7-16	0470	7-58
0378	8-70	0409	7-46	0440	7-17	0471	7-59
0379	8-76	0410	7-44	0441	4-55	0472	7-87
0380	8-73	0411	7-32	0442	4-56	0473	7-88
0381	8-74	0412	7-37	0443	7-6	0474	7-52
0382	8-71	0413	7-40	0444	7-10	0475	7-60
0383	8-82	0414	7-45	0445	7-9	0476	7-89
0384	8-84	0415	7-36	0446	7-54	0477	7-72
0385	8-83	0416	7-35	0447	7-55	0478	7-86
0386	8-85	0417	7-31	0448	7-56	0479	7-51
0387	8-1	0418	7-30	0449	7-57	0480	7-76
0388	8-86	0419	7-33	0450	7-73	0481	7-77
0389	8-87	0420	7-34	0451	7-61	0482	7-78
0390	8-89	0421	7-43	0452	2-12	0483	7-79
0391	8-88	0422	7-41	0453	7-24	0484	7-80
0392	8-90	0423	7-38	0454	2-13	0485	7-81
0393	8-2	0424	7-84	0455	7-7	0486	7-82
0394	8-91	0425	7-85	0456	7-8	0487	7-83
0395	7-53	0426	7-67	0457	7-14	0488	10-1
0396	7-48	0427	7-68	0458	7-11	0489	10-3
0397	7-49	0428	7-69	0459	7-13	0490	10-2
0398	7-47	0429	7-70	0460	10-22	0491	10-6
0399	7-50	0430	7-71	0461	7-27	0492	10-7
0400	7-1	0431	7-42	0462	7-29	0493	10-4
0401	3-1	0432	7-4	0463	7-28	0494	10-5
0402	7-2	0433	7-3	0464	7-5	0495	9-2
0403	7-23	0434	7-39	0465	7-62	0496	9-3
0404	7-21	0435	7-15	0466	7-63	0497	9-4
0405	7-22	0436	7-12	0467	7-64	0498	3-43
0406	7-26	0437	7-19	0468	7-65	0499	7-75
0407	7-18	0438	7-20	0469	7-66	0500	7-74

AUTHOR'S RECOMMENDATION

The Experimental Aircraft Association, Inc. is a very successful and effective nonprofit organization that represents and serves those of us interested in flying, in general, and in sport aviation, in particular. I personally invite you to enjoy becoming a member:

    $30 for a one-year membership
    $18 per year for individuals under 19 years old
    Family membership available for an extra $10 per year.

Membership includes the monthly magazine *Sport Aviation*.

Write to: Experimental Aircraft Association, Inc.
                Wittman Airfield
                Oshkosh, WI 54903

or call: (414) 426-4800
                (800) 322-2412   (in Wisconsin: 1-800-236-4800)

The annual EEA Oshkosh Fly-In is an unbelievable aviation spectacular with over 10,000 airplanes at one airport! Virtually everything aviation-oriented you can imagine! Plan to spend at least one day (not everything can be seen in a day) in Oshkosh (100 miles northwest of Milwaukee).

Convention dates: 1990 - July 27 through August 3
                         1991 - July 26 through August 1
                         1992 - July 31 through August 6

INSTRUCTOR CERTIFICATION FORM
WRITTEN TEST

Name: _____

   I certify that I have reviewed the above individual's completion of the *RECREATIONAL PILOT FAA WRITTEN EXAM* home-study course by Irvin N. Gleim for the FAA Recreational Pilot written test (covering the topics specified in FAR 61.97). I find that (s)he has satisfactorily completed the course and find him/her competent to pass the recreational pilot (airplane) written test.

Signed: _____

CFI Number and
Expiration Date: _____

WRITTEN EXAM BOOKS
available from
AVIATION PUBLICATIONS, Inc.

RECREATIONAL PILOT FAA WRITTEN EXAM ($9.95)
The newest certificate offered by the FAA. The FAA's written test consists of 50 questions out of the 445 questions in our book.

PRIVATE PILOT FAA WRITTEN EXAM ($9.95)
The FAA's written test consists of 50 questions out of the 673 questions in our book.

INSTRUMENT PILOT FAA WRITTEN EXAM ($14.95)
The FAA's written test consists of 60 questions out of the 834 questions in our book. Also, those people who wish to become an instrument-rated flight instructor (CFII) or an instrument ground instructor (IGI) must take the FAA's written test of 50 questions from this book.

COMMERCIAL PILOT FAA WRITTEN EXAM ($14.95)
The FAA's written test consists of 100 questions out of the 575 questions in our book.

FUNDAMENTALS OF INSTRUCTING FAA WRITTEN EXAM ($9.95)
The FAA's written test consists of 50 questions out of the 182 questions in our book. This is required of any person to become a flight instructor or ground instructor. The test only needs to be taken once, so if someone is already a flight instructor and wants to become a ground instructor, taking the FOI test a second time is not required. Conversely, a ground instructor who wants to become a flight instructor does not have to take the FOI test again.

FLIGHT/GROUND INSTRUCTOR FAA WRITTEN EXAM ($14.95)
The FAA's written test consists of 100 questions out of the 834 questions in our book. To be used for the Certificated Flight Instructor (CFI) written test and those who aspire to the Advanced Ground Instructor (AGI) rating for airplanes. Note that this book also covers what is known as the Basic Ground Instructor (BGI) rating. However, the BGI is not useful because it does not give the holder full authority to sign off private pilots to take their written test. In other words, this book should be used for the AGI rating.

FLIGHT MANEUVER AND REFERENCE BOOKS
available from
AVIATION PUBLICATIONS, Inc.

RECREATIONAL PILOT FLIGHT MANEUVERS ($11.95)

Contains, in outline format, pertinent information necessary to be a skilled recreational pilot. An excellent reference book to begin your flight training endeavors with.

PRIVATE PILOT HANDBOOK ($11.95)

A complete private pilot ground school text in outline format with many diagrams for ease in understanding. To make it more useful and save user's time, it has a very complete, detailed index. It contains a special section on biennial flight reviews.

PRIVATE PILOT FLIGHT MANEUVERS ($9.95)

A complete guide for airplane flight training leading to the private pilot certificate. Covers the flight maneuvers in outline format with numerous diagrams. It includes a reprint of the new FAA Practical Test Standards, with outlines of the maneuvers to be performed on the FAA airplane flight test (including acceptable performance limits).

INSTRUMENT PILOT FLIGHT MANEUVERS & HANDBOOK ($17.95)

A comprehensive explanation with illustrations of flight by reference to instruments, IFR charts and navigation, instrument approaches, IFR cross-country, and includes complete discussion of the current FAA Practical Test Standards for the instrument rating-airplane.

MULTIENGINE AND SEAPLANE FLIGHT MANEUVERS & HANDBOOK ($14.95)

Provides separate and comprehensive explanation of each rating. Emphasis is placed on how each airplane is different from single-engine airplanes. Flight and ground operation is explained, illustrated, and analyzed in terms of the FAA Practical Test Standards. This book is extremely helpful as a reference book as well as a training text.

COMMERCIAL PILOT AND FLIGHT INSTRUCTOR FLIGHT MANEUVERS & HANDBOOK ($14.95)

Separate and complete coverage of the commercial certificate and flight instructor certificate. It begins with an explanation of the requirements and recommended steps to obtain each certificate. All flight maneuvers are explained, illustrated, and analyzed in terms of the current FAA Practical Test Standards. It is an excellent reference text for professional pilots.

MAIL TO:	**AVIATION PUBLICATIONS, Inc.** P.O. Box 12848 University Station Gainesville, FL 32604
OR CALL:	(904) 375-0772

Books Currently Available

RECREATIONAL PILOT FAA WRITTEN EXAM	First (1989-1991) Edition	$ 9.95	$ _____
PRIVATE PILOT FAA WRITTEN EXAM	Fourth (1988-1990) Edition	9.95	_____
INSTRUMENT PILOT FAA WRITTEN EXAM	Second (1988-1990) Edition	14.95	_____
COMMERCIAL PILOT FAA WRITTEN EXAM	Second (1988-1990) Edition	14.95	_____
FUNDAMENTALS OF INSTRUCTING FAA WRITTEN EXAM	Third (1989-1991) Edition	9.95	_____
FLIGHT/GROUND INSTRUCTOR FAA WRITTEN EXAM	Third (1989-1991) Edition	14.95	_____
RECREATIONAL PILOT FLIGHT MANEUVERS (1st Edition)		$11.95	_____
PRIVATE PILOT HANDBOOK (3rd Edition)		11.95	_____
PRIVATE PILOT FLIGHT MANEUVERS (2nd Edition)		9.95	_____
INSTRUMENT PILOT FLIGHT MANEUVERS & HANDBOOK (1st Edition)		17.95	_____
MULTIENGINE AND SEAPLANE FLIGHT MANEUVERS & HANDBOOK (1st Edition)		14.95	_____
COMMERCIAL PILOT AND FLIGHT INSTRUCTOR FLIGHT MANEUVERS & HANDBOOK (1st Edition)		14.95	_____

Florida residents add 6% sales tax .... 6% Tax _____
Please call or write for additional charges for out-of-the-U.S. shipments

TOTAL $ _____

1. We process and ship orders within one day of receipt of your order. We generally ship via UPS for the Eastern U.S. and U.S. mail for the Western U.S.

2. Please PHOTOCOPY this order form for friends and others.

3. No CODs. All orders from individuals must be prepaid and are protected by our unequivocal refund policy.
   Library and company orders may be on account.
   Shipping charges will be added to telephone orders and to orders not prepaid.

Name _____

Mailing Address _____

Street Address (for UPS) _____

City _____ State _____ Zip _____

☐ MasterCard ☐ VISA ☐ Check or Money Order Enclosed

MasterCard/VISA No.

___ ___ ___ ___ - ___ ___ ___ ___ - ___ ___ ___ ___ - ___ ___ ___ ___

Expiration Date *(month/year)* ___/___

Signature _____

050

AVIATION PUBLICATIONS GUARANTEES
AN IMMEDIATE, COMPLETE REFUND
ON ALL MAIL ORDERS
IF A RESALABLE TEXT IS RETURNED IN 30 DAYS

Additional Order Form

MAIL TO: **AVIATION PUBLICATIONS, Inc.**
P.O. Box 12848
University Station
Gainesville, FL 32604

OR CALL: **(904) 375-0772**

Books Currently Available

RECREATIONAL PILOT FAA WRITTEN EXAM	First (1989-1991) Edition	$ 9.95	$ _____
PRIVATE PILOT FAA WRITTEN EXAM	Fourth (1988-1990) Edition	9.95	_____
INSTRUMENT PILOT FAA WRITTEN EXAM	Second (1988-1990) Edition	14.95	_____
COMMERCIAL PILOT FAA WRITTEN EXAM	Second (1988-1990) Edition	14.95	_____
FUNDAMENTALS OF INSTRUCTING FAA WRITTEN EXAM	Third (1989-1991) Edition	9.95	_____
FLIGHT/GROUND INSTRUCTOR FAA WRITTEN EXAM	Third (1989-1991) Edition	14.95	_____
RECREATIONAL PILOT FLIGHT MANEUVERS (First Edition)		$11.95	_____
PRIVATE PILOT HANDBOOK (Third Edition)		11.95	_____
PRIVATE PILOT FLIGHT MANEUVERS (Second Edition)		9.95	_____
INSTRUMENT PILOT FLIGHT MANEUVERS & HANDBOOK (First Edition)		17.95	_____
MULTIENGINE AND SEAPLANE FLIGHT MANEUVERS & HANDBOOK (First Edition)		14.95	_____
COMMERCIAL PILOT AND FLIGHT INSTRUCTOR FLIGHT MANEUVERS & HANDBOOK (First Edition)		14.95	_____

Florida residents add 6% sales tax .... 6% Tax _____

Please call or write for additional charges for out-of-the-U.S. shipments

                                                                            TOTAL $ _____

1. We process and ship orders within one day of receipt of your order. We generally ship via UPS for the Eastern U.S. and U.S. mail for the Western U.S.

2. Please PHOTOCOPY this order form for friends and others.

3. No CODs. All orders from individuals must be prepaid and are protected by our unequivocal refund policy. Library and company orders may be on account. Shipping charges will be added to telephone orders and to orders not prepaid.

Name _____

Mailing Address _____

Street Address (for UPS) _____

City _____ State _____ Zip _____

☐ MasterCard   ☐ VISA   ☐ Check or Money Order Enclosed

MasterCard/VISA No.

___ ___ ___ ___ - ___ ___ ___ ___ - ___ ___ ___ ___ - ___ ___ ___ ___

Expiration Date *(month/year)* ____/____

Signature _____

050

AVIATION PUBLICATIONS GUARANTEES

AN IMMEDIATE, COMPLETE REFUND

ON ALL MAIL ORDERS

IF A RESALABLE TEXT IS RETURNED IN 30 DAYS

**Additional Order Form
is on reverse side**

INDEX

Abnormal combustion - 53, 70
Absolute altitude - 51
Acceleration/deceleration errors - 49
Advisory circulars, FAA - 171, 175
Aerobatic flight - 143, 162
Aerodynamic forces - 15, 20
Aeronautical charts - 172, 178
AFD - 171, 176
AIM - 172, 176
Air mass thunderstorms - 113
Aircraft speed - 142, 162
Airman written test
    Application - 8
    Report - 12
Airmen, AC 60 - 171
AIRMETs - 109, 135
Airplane
    Stability - 17, 24
    Turn - 17, 24
Airport
    Advisory areas - 76, 85
    Elevation - 172
    Identifiers - 174
    Location - 172
    Radar service areas - 76, 85, 173
    Traffic areas - 76, 84
    Traffic patterns - 74, 79
Airport/Facility Directory - 171, 176
Airspace - 140, 143
Airspace, AC 70 - 171
Airspeed, flight operation - 186, 193
Airspeed indicator - 50, 55
Airworthiness, civil aircraft - 141, 159
Airworthiness certificate - 138, 147
Alcohol and drugs - 140, 157
Alert areas - 172, 174
Altimeter - 50, 59
    Calculations - 51, 62
    Errors - 51, 62
Altitude
    Cruising - 145, 167
    Definitions - 51
    Minimum - 143, 163
    Types - 50, 61
Altocumulus castellanus - 118
Angle of attack - 16, 20
Anvil, thunderstorm - 112
Application, airman written test - 8
Area weather forecasts - 106, 124

ARSAs - 76, 85, 173
Asymmetric propeller loading - 17
ATC and general operations, AC 90 - 171
Attack, angle - 16, 20
Authorization for test - 7
Aviation, compass - 49
Aviation fuel practices - 53, 71

Barometric pressure/density altitude - 30
Basic empty weight - 87
Beacons - 75, 81
Best-angle-of-climb speed - 187
Best-rate-of-climb speed - 187
BFR - 138
Biennial flight review - 138
Briefings, weather - 105, 122

Calculations, altimeter - 51, 62
Camber, wing - 21
Carbon dioxide - 185
Carbon monoxide - 186, 189
Carburetor heat - 53, 69
Carburetor icing - 52, 67
Causes of weather - 101, 110
Center of
    Gravity - 87
    Gravity moment envelope - 93
    Lift - 17
    Pressure - 17
Centrifugal force - 24
Certificates and ratings - 138, 147
Certifications, aircraft - 141, 159
CG - 87
CG limit, rear - 17
Change of address - 139, 150
Checklists - 186
Chevron, on runway - 74
Chord line - 16
Civil aircraft airworthiness - 141, 159
Civil aircraft certifications - 141, 159
Clearing turns - 194
Cloud heights - 104
Clouds - 103, 117
Collision avoidance - 178, 186, 194
Combustion, abnormal - 53, 70
Compass turning errors - 49, 54
Constant-speed propeller - 52, 65

213

Continuous participation - 104
Control zones - 140, 155
Controlled airspace - 173
Coriolis force - 101
Crosswind components - 32, 42
Cruise performance - 32, 40
Cruising altitudes, VFR - 145, 167
Cumulonimbus clouds - 103
Cumulus stage, thunderstorm - 102

Density altitude - 29, 34, 51
    Barometric pressure - 30
    Chart - 36
    Computations - 30, 36
Depiction charts, weather - 107, 126
Detonation - 53
Dewpoint - 103, 115
Dihedral, wing - 21
Disorientation, spatial - 185, 189
Displaced threshold, runway - 73
Dissipating stage, thunderstorm - 102
Drag - 15
Drugs and alcohol - 140, 157

Elevation, airport - 172
ELTs - 142, 160
Embedded thunderstorms - 102
Emergency locator transmitters - 142, 160
Empty weight, airplane - 87
Engine
    Ignition system - 52, 66
    Starting - 186, 191
    Temperature - 52, 64
En Route advisory service - 108, 129
Errors, altimeter - 51, 62
Evaporation - 103
Examination procedures - 11

FAA
    Advisory circulars - 171, 175
    Form 8080-2 - 12
    Question numbers - 13
    Test instructions - 14
    Written test - 4
Failure on written test - 12
FARs - 137
Federal airways - 140, 155
Federal Aviation Regulations - 137
Flaps - 15, 19

Flight
    Manuals, markings, and placards - 141, 159
    Operations - 185
    Physiology - 185
    Review - 138, 148
    Standards District Offices - 6
Floating, on landing - 24
Float-type carburetor - 67
Foehn gap - 110
Fog - 103, 115
Form 8080-2 - 12
Format, written test - 7
Freezing rain - 102
Frontogenesis - 110
Frontolysis - 110
Fronts, weather - 102, 110
Frost - 16, 22
FSDOs - 6
Fuel
    Air mixture - 53, 69
    Ignition system - 52
    Practices - 53, 71
    Requirements, VFR - 141, 158
    Servicing, airport - 172

G units - 18
Geographic position airport - 172
Green arc, airspeed - 50
Ground effect - 16, 23

Handpropping - 186
Heat, carburetor - 53, 69
Horizontal component of lift - 17
Hyperventilation - 185, 188

Ice pellets - 102
Icing - 102, 113
Icing, carburetor - 52, 67
Ignition system - 52, 66
Incidence, wing - 21
Indicated altitude - 51
Induced drag - 17
Inspection,
    Maintenance - 145, 168
    Preflight - 186, 190
Instruction, written test - 11
Instrument approach procedures - 172
Invection fog - 103
Inversions, temperature - 104, 121

Landing
    Direction indicator - 74
    Distance - 33, 44
    Strip indicator - 74
Lapse rate - 104
Latitude - 172
Lecticular clouds - 118
Left-turning tendency - 17
Lift - 15
    Center of - 17
Load factor - 18, 26
Load factor chart - 18
Loading graph - 93
Loading graph, CG - 88
Longitude - 172
Longitudinal stability - 17

Magnetic course - 173
Maintenance
    Records - 145, 169
    Requirements - 145, 168
Maneuvering speed - 50, 186
Mature stage, thunderstorm - 102
Maximum
    Flaps-extended speed - 56
    Gross weight - 88
    Structural cruising speed - 56
Medical certificate - 2, 138, 148
Military flight routes - 174
Military operation areas - 172
Minimum safe altitudes - 143, 163
MOA - 172
Moments, CG - 90
MULTICOM - 75, 82

National Transportation Safety Board - 146, 169
Never-exceed speed - 56
Nimbostratus - 118
Nimbus clouds - 103
Nose-down pitch - 25
NOTAM service - 172
Notification of written test score - 11
NTSB - 146, 169

Obstructions on sectional charts - 174
Occlusion - 113
Optional equipment weights - 87
Overbanking tendency - 19
Overrun, runway - 74
Oxygen availability - 172

P-factor - 17, 26
Parachuting - 141, 158
Performance, cruise - 32, 40
Pilot-in-command responsibility - 140, 156
Pilot reports - 133
PIREPs - 133
Pitch, nose-down - 25
Power failure - 193
Preflight - 140, 156
    Inspection - 186, 190
Prefrontal system - 113
Preignition - 53
Preparing for written test - 5
Pressure altitude - 30, 51
Pressure, center of - 17
Procedures, examination - 11
Prognostic charts, low-level - 108, 131
Prohibited areas - 143, 164
Propeller, constant-speed - 52, 65
Propping the engine - 186

Question numbers, FAA - 13
Question selection sheet - 9

Radar summary charts - 108, 129
Radiation fog - 103
Ratings and certificates - 138, 147
Rear CG limit - 17
Recency of flight experience - 139, 149
Recreational pilot certificate - 2, 139, 150
Redline, airspeed - 50
Reference datum - 87
Relative humidity changes/density altitude - 30
Relative wind - 16
Restricted areas - 143, 164
Restricted category aircraft - 142, 160
Right-of-way rules - 142, 160
Roll cloud - 111
Rotating light beacon - 172
Rudder - 15, 19
Runway markings - 73, 77
Runway numbering system - 78

Safe airspeed - 187
Safety belts - 140, 157
Seabreezes - 102
Sectional charts - 172, 178
Segmented circle - 74
Selection sheet, question - 9
Severe turbulence - 193

SIGMETs - 109, 135
Spatial disorientation - 185, 189
Speed, aircraft - 142, 162
Spins - 16, 21
Squall line - 102
Stability - 17, 24
    Of air masses - 104, 119
    Weather - 104, 121
Stalls - 16, 21
Starting engine - 186, 191
Stationary front - 114
Steady state thunderstorms - 113
Stratoform clouds - 104
Structural icing - 102
Student pilot certificate - 3
Sublimation - 16
Surface weather reports - 105, 122

Takeoff distance - 30, 31
Taxiing - 186, 191
TCAs - 76, 85, 143, 164
Temperature
    Changes/density alt - 30
    Dewpoint - 103, 115
    Engine - 52, 64
    Inversions - 104, 121
Terminal control areas - 76, 85, 143, 164
Terminal forecasts - 106, 124
Test instructions, FAA - 14
Threshold, runway - 73
Thrust - 15
Thunderstorms - 102, 111
Time conversion - 172
Torque - 17, 26
Towering cumulus clouds - 103
Traffic pattern altitude - 172
Traffic patterns - 74, 79
Transcribed weather broadcasts - 108, 133
Tri-colored VASI - 75
True altitude - 51
True vs. indicated altitude - 63
Turbulent air - 50
Turning errors, compass - 49, 54
Turns - 17, 24
TWEBs - 108, 133

UNICOM - 75, 82
Unusable fuel - 87

VASI - 75, 82
Vertigo - 185
VFR cruising altitudes - 145, 167
VFR weather minimums - 144, 165
Visual approach slope indicators - 75, 82
Vortices - 33, 47

Weather - 101
    Briefings - 105, 122
    Causes - 101, 110
    Depiction charts - 107, 126
    Minimums, VFR - 144, 165
    Reports - 105, 122
Weight - 15
    And balance - 87
    And balance tables - 96
What to take to written test - 11
When to take test - 6
Where to take test - 6
White arc, airspeed - 50
Wildlife refuge, minimum altitude - 172
Wind
    And temperature aloft forecast - 109, 134
    Direction indicator - 74
    Shear - 103, 114
Wing camber - 19
Wing chord line - 16
Wingtip vortices - 33, 47
Written test - 4
    Instructions - 11
    Score notification - 11

Yellow arc, airspeed - 50

Recreational Pilot FAA Written Exam, First (1989-1991) Edition, First Printing

Please forward your suggestions, corrections, and comments concerning typographical errors, etc. to **Irvin N. Gleim** • c/o Aviation Publications, Inc. • **P.O. Box 12848** • **University Station** • **Gainesville, Florida** • **32604**. Please include your name and address so we can properly thank you for your interest. Also, please refer to both the page number and the FAA question number for each item.

1. _____

2. _____

3. _____

4. _____

5. _____

6. _____

7. _____

8. _____

9. _____

10. _____

Name: _____

Address: _____

_____

City/State/Zip: _____

Telephone: (_____) _____